# HISTORICAL AND DESCRIPTIVE NARRATIVE
## OF THE
## MAMMOTH CAVE OF KENTUCKY.

### INCLUDING

### EXPLANATIONS OF THE CAUSES CONCERNED IN ITS FORMATION, ITS ATMOSPHERIC CONDITIONS, ITS CHEMISTRY, GEOLOGY, ZOOLOGY, ETC.
### WITH
### FULL SCIENTIFIC DETAILS OF THE EYELESS FISHES.

BY

## W. STUMP FORWOOD, M.D.,

This work

is

RESPECTFULLY INSCRIBED

TO

JAY COOKE, JR.

ONE OF THE COMPANIONS OF OUR JOURNEY,

AND A LIBERAL PATRON OF LITERATURE,

BY HIS FRIEND,

THE AUTHOR.

PHILADELPHIA, APRIL, 1870.

Table of Contents

# CHAPTER I. - INTRODUCTION.

It is our purpose to describe, from our own observations made in the spring of 1867, and from the observations of others, that grand and weird cavern known as the Mammoth Cave of Kentucky, a wonder of its kind, unequaled in America or in the world, within whose sublime portals travelers have confessed the most profound awe at entrance, and the greatest rapture when its glorious mysteries were made visible to them.

We did not make the visit with the view of informing the public what was to be seen, but simply for the purpose of gratifying our individual curiosity.

Finding the object to be one of greater magnitude than was anticipated, it occurred to us, as an afterthought, that a short sketch might interest a friend at home. In executing this intention, it was soon discovered that a surprising number of pages were required to give even a brief intelligible outline of the great cavern.

It was then suggested that the sketch which had been commenced should be extended, and published in book form, that the information it contained might be accessible to the general public, instead of being restricted to one or two friends, as at first designed. This suggestion, though not consonant with our feelings when first proposed, has, upon reflection, been adopted.

Desiring to obtain some profitable information in advance of our visit to the Cave, we applied successively to the principal booksellers in Boston,

New York, Philadelphia, Baltimore, and Cincinnati for the purchase of a descriptive work, and were greatly surprised and disappointed by the answer in each case, that not one of them had any publication on the subject, neither had they any knowledge of the existence of such a work. This deficiency in the book market appeared to us extraordinary, for it is presumable that all persons of any education in this country, and many abroad, have heard of the existence of the Cave, and are aware that it is a curiosity of more than ordinary importance; it is therefore a matter of astonishment that no general account of it can be obtained among the booksellers by those who are desirous of information regarding its wonders.

Upon arriving at the Cave we found a small pamphlet for sale, entitled "A Guide Manual to the Mammoth Cave of Kentucky. By Charles W. Wright, M.D., Professor of Chemistry in the Kentucky School of Medicine, formerly Professor of Chemistry in the Medical College of Ohio." Printed at Louisville, Ky., by Bradley Sr, Gilbert, 1860.

This manual explains very satisfactorily the chemical and mechanical causes which were exerted in the formation of the Cave, and briefly indicates all the chief points of interest which should attract the notice of the visitor in his explorations, and, we believe, is thoroughly reliable in all these particulars. Its circulation, however, is limited to those who visit the Cave, rarely coming before the general reader; and it is probable that the majority of visitors, as in our case, have no opportunity of examining and profiting by it until after their departure, and then it is generally thrown aside and forgotten.

Since our visit we have made every effort to procure all that has been written on the subject, with the hope of thus making our account as complete as possible. We have, therefore, delayed its publication for nearly three years.

We have succeeded in obtaining four short articles, chiefly scientific, in as many different numbers of Silliman's "American Journal of Science and Arts," written by Professors Lock, Agassiz, Silliman, and Wyman, the first dating back as far as 1842; also a rather lengthy description given by our great American traveler, Bayard Taylor, who charmingly invests every sketch of Nature's works touched by his pen with the glowing light of romance, so appropriate in this case.

We have also found a copy of a manual called "Pictorial Guide to the Mammoth Cave, Kentucky. By the Rev. Horace Martin." New York: Stringer & Townsend, 1851; with ten illustrations, pp. (including blanks for notes) 116, long out of print. A brief article on the Cave, in a book entitled "The Hundred Wonders of the World," has recently been brought to our notice; also an article in Collins's "History of Kentucky" (1847), and a few pages in Dr. Poucher's work on the Universe, etc., translated from the French, 1870. The pamphlet of Dr. Binkerd ("The Mammoth Cave and its Denizens: A Complete Descriptive Guide. By A. D. Binkerd, M.D." Cincinnati: Robert Clarke & Co., Printers, 1839. Pamphlet, pp. 96) has appeared since the greater part of the present work has been in manuscript; and in completing our materials for the press we are unable to derive any assistance from the work of Dr. Binkerd.

Several newspaper articles, worthy of but little attention, have also come under our notice.

At the eleventh hour, since our manuscript was placed in the hands of the printers, we have succeeded, through the kindness of Mr. John R. Proctor, of Maysville, Ky., in procuring a copy of a work entitled "Rambles in the Mammoth Cave during the Year 1841. By a Visitor." Louisville, Ky.: Morton & Griswold, 1845.

This list comprises all the works on the subject that we have any knowledge of with the exception of one by Mr. Lee, Civil Engineer, published about the year 1840, said to be of some value, but the most diligent search on our part has proved unsuccessful in finding a single copy.

In preparing this history of the Mammoth Cave, we make as much use as possible of the materials just mentioned, collating their agreements and disagreements with our own observations. We are chiefly indebted, however, to the valuable Manual of Professor Wright for all measurements and material facts, such as can be acquired only by a protracted series of observations; and we trust that this general announcement of the authorities that we draw upon will serve us in many instances instead of quotation marks.

# CHAPTER II. - THE CAVE.

In order that the reader may form, at the outset, some idea of the general outline and physical character of the Mammoth Cave, we will ask him to imagine the channel of a large and winding river, with tributaries at intervals, some of them the size of the main stream, emptying into the chief river, like for instance, the Missouri and Ohio joining the Mississippi; these tributaries also receiving their support from creeks, branches, and rivulets, some of them quite small and extending but a short distance, while others are much larger, longer, and more beautiful. Now, it is easy to imagine these rivers as being under ground, or having a surface covering of earth and rock, and that their rugged channels and banks have ceased, from some cause, to be bathed with the waters which, in ages long past, flowed so freely along them; in fact, that they are perfectly dry, except in a few of the avenues.

By the aid of this illustration it may also be comprehended why so much travel is necessary, as will be presently stated, to visit the different parts of the Cave. We are obliged to follow each tributary of the chief river to its source, and to return by the same route to its mouth, at the point of our departure; thus duplicating the distances of all the rivers, creeks, etc.

It is exceedingly difficult to obtain information regarding the early history of the Cave, simply from the fact that it was not explored to any great extent for several years after its discovery, and that the early explorers did not regard it as a curiosity of sufficient importance to call for the publication of any detailed account. It has been stated by Bayard Taylor, and others, that the discovery of the Cave dated back as far as the year 1802; but we are fortunate in possessing a highly interesting and valuable letter from Mr. Frank Gorin, a former. proprietor of the Cave, addressed to the author some months after his visit, and with permission, hereto appended in full, which fixes the date of the discovery in the year 1809. The letter contains several facts that will here anticipate their regular order, and will again be adverted to:

"GLASGOW, KY., Feb. 9, 1868.
"W. STUMP FORWOOD, M.D
"DEAR SIR:
"I am in receipt of yours of the 27th ultimo. You desire all the information I can give respecting the date of the discovery, the early history, the operations of the saltpetre miners, etc. of the Mammoth Cave.
"This part of Kentucky was peopled and settled in the latter part of the eighteenth and early part of the nineteenth century. The Mammoth Cave is situated on the south side of Green River, and not far from its banks. It was discovered in the year 1809, by a man named Houchins, by running a bear into it. The entrance was small, although there was a large 'sink' at the mouth. This is not the original mouth or entrance: the original mouth is about one fourth of a mile north, or northwest, from the present entrance. It is a deep hole, perhaps fifty feet across at the top, and was

doubtless the site, years, years ago, of one of those large springs so often found near the south bank of Green River. There is a spring at the present entrance of the Cave; the water, no doubt, caused the falling of the· roof, and closed up at that place the channel leading from the former mouth.

"Very few persons know of the original 'mouth,' as the Cave at its present mouth is filled up with rocks, dirt, etc.

"When first discovered, the Cave was not considered of much value. It sold, with about two hundred acres of land, for about forty dollars.

"McLean, I believe, was the first person who attempted to make saltpetre there, perhaps in the year 1811. He sold to Gatewood, who enlarged the works. He, in turn, sold to Gratz & Wilkins (Gratz of Philadelphia, and Wilkins of Lexington, Ky.); they, during the War of 1812, made large quantities of saltpetre, and wagoned it principally to Philadelphia. Their agent at the Cave was an Irish gentleman by the name of Archibald Miller. The work during the War of 1812 was mostly done by negroes, some of them working in the Cave without coming out for an entire year. They came out healthy, and had a beautiful gloss, with shining faces and skins. This is singular, as they were rarely ever more than one mile from the entrance.

"After the War of 1812-14 it was no longer profitable to make saltpetre at the Cave, on account of the importation of the East Indian article in the Eastern market, at rates much cheaper than it could be wagoned from the Cave.

"When Messrs. Gratz & Wilkins ceased to make saltpetre, after having acquired sixteen hundred and ten acres of land over and around the Cave, they continued their faithful, true, and honest agent, Miller, to overlook and take care of the property and to show the Cave to the curious. About the year 1816, Mr. Miller placed the Cave and other property in the possession of his brother-in-law, Mr. Moore, and his wife, both Irish, of the old stock. Mr. Moore had been wealthy, and a large merchant in Philadelphia. Unfortunately, he was seduced into unlawful acts by Blennerhassett, the friend of Burr, and was pecuniarily ruined. The Moores left there some time afterward, when Gatewood took possession, and showed the Cave to all visitors for years; but it did not pay, and he left.

"In 1837 I purchased the Cave and property, when it was in a dilapidated state, and placed Mrs. Moore there (Mr. Moore having previously died), together with Archibald Miller, her nephew, and son of the previous occupant of the same name, as my agents. They were residing there when I sold the Cave and property to Dr. John Crogan, who continued Mrs. Moore and Mr. Miller, Jr. in charge during their lives. Dr. Grogan devised the estate to Mr. Gwathmey and Judge J. R. Underwood, for the use of eleven nephews and nieces. Judge Underwood is the surviving trustee, and is now managing the estate.

(We remember having seen a statement in the newspapers, years ago, to the effect that Dr. Crogan, while visiting objects of interest in Europe, was repeatedly asked for information regarding the Mammoth Cave; and, as the result of the mortification induced by his total ignorance of the subject, on his return home he visited the curiosity, and purchased the property, with the view of imparting more extended knowledge of this great American wonder to his countrymen and to travelers from other lands.)

"It was while I owned the property that a nephew of mine, Mr. Charles F. Harvey (now a merchant in Louisville, Ky.), was lost in the Cave for thirty nine hours. After he was found, I determined to have further explorations made. At that time no person had ever been beyond the 'Bottomless Pit.' We discovered 'Gorin's Dome,' covering nearly an acre of ground, and perhaps five hundred feet high. (Mr. Gorin, in a subsequent letter to the writer, states that possibly this estimate of the dimensions of the Dome is too great, as our own observations confirm; but he believes that Dr. Wright's estimate, which we will hereafter give, is much below the actual measurements.—W. S. F.)

"I placed a guide in the Cave, the *celebrated* and *great* STEPHEN, and he aided in making the discoveries. He was the first person who ever crossed the 'Bottomless Pit'; and he, myself, and another person, whose name I have forgotten, were the only persons ever at the bottom of 'Gorin's Dome,' to my knowledge.

"After Stephen crossed the 'Bottomless Pit,' we discovered all that part of the Cave now known beyond that point. Previous to those discoveries, all interest centered in what is known as the 'Old Cave,' the chief points of attraction being the 'Star Chamber,' the 'Cataract,' 'The Chief City,'

'Robber's Cave,' 'Lover's Leap,' 'Bonaparte's Breastworks,' 'Gatewood's Dining Table,' 'Black Chambers,' 'Grotto,' etc. etc.; but now many of these points are but little known, although, as Stephen was wont to say, they were 'grand, gloomy, and peculiar.'

"Many attempted descriptions of the Cave have been published in the newspapers; and several pamphlet publications have been made; but I know of none now existing. Many of the newspaper articles were utterly false.

"Stephen was a self educated man; he had a fine genius, a great fund of wit and humor, some little knowledge of Latin and Greek, and much knowledge of geology; but his great talent was a perfect knowledge of man. (It has been said that Stephen was partly of Indian extraction. In reply to a subsequent letter addressed to Mr. Gorin, on this and other points, he remarks, "There was not any Indian blood in Stephen's veins. I knew his reputed father, who was a white man. I owned Stephen's mother and brother, but not until after both of the children were born. Stephen was certainly a very extraordinary boy and man. His talents were of the first order. He was trustworthy and reliable; he was companionable; he was a hero, and could be a clown. He knew a gentleman or a lady as if by instinct. He learned whatever he wished, without trouble or labor; and it is said that a late professor of geology spoke highly of his knowledge in that department of science.")

"I have been compelled to write you this letter in great haste, but you may rely upon the facts as stated.

"Yours truly,

"F. GORIN."

From data that we have obtained from various sources, we learn that the 'Bottomless Pit' was not crossed, nor the great curiosities beyond dreamed of, for about thirty years after what is called the 'Main Cave' had been explored. Indeed, it is known that many avenues, with their hidden treasures, have not to the present day been trodden by mortal footsteps. So much has already been explored that curiosity appears to be satiated.

It is said that about one hundred and fifty miles of travel is required to visit the parts of the Cave that have already been traversed; and we were informed by the guides that avenues were known to them which would probably increase the extent of travel to two hundred miles. (Since the foregoing was penned, we have been informed by the proprietor of the Cave Hotel, Mr. L. J. Proctor, in a letter dated March 12, 1870, that "Two years ago three of the guides at the Cave, Messrs. F. M. De Monbrum and Charles and A. Merideth, discovered a new avenue in the Mammoth Cave, branching off from the Pass of El Ghor, just beyond Ole Bull's Concert Room. They first entered a narrow crevice, through which they passed some seventy yards, when they entered a large cave, which they explored for many miles, and from which many branching avenues led off, which they did not explore. They describe this newly discovered avenue as extremely grand, and in many places beautiful.

They crossed a large, and as yet unexplored river and found that the main avenue terminated in a dome more extensive than any that they had ever seen. What was beyond this dome they could not conjecture, as they were unable to enter it from the avenue. They estimate that they traveled eight miles in this one avenue. I have not seen it myself. The explored portions of the Cave that I have visited constitute within themselves an underground world; and I am satisfied that I have traveled from 150 to 200 miles in the different avenues, upon the Long Route especially. There is a perfect wilderness of Cave that is never seen or dreamed of by visitors generally, and many parts more beautiful than those ordinarily seen by parties making the Long Route. I refer particularly to Marion's Avenue, Alida's Avenue, Murdock's Pass, and out Boone's Avenue and the regions of Mystic River."

# CHAPTER III. - LOCATION OF THE CAVE.

The Mammoth Cave is situated in Edmonson County, in the southern part of Kentucky. It is most readily approached from the North by way of Louisville, by the Louisville & Nashville Railroad, which has long since superseded the old stage coaches. The distance from Louisville is about ninety five miles, or about one half the distance between that city and Nashville.

The station at which passengers left the railroad at the time of our visit is called Cave City, a point about ten miles from the Cave.

Visitors from the South come by way of Nashville to the same point. The high sounding name of "City," as applied to this place, reminded us forcibly of the vest of the hero of the comic song, which, he said "was big enough for two; But there is nothing strange in that; For the tailor saw, without a doubt, I some day would grow fat!"

This "City" consists of about a dozen ordinary looking houses; but, possessing an ample title in advance, it may be presumed that it will some day grow large.

The hotel from which the stage coach line starts is small, but the traveler is very comfortably entertained. (Since the above was written, we regret to learn that this little city was, on January 17, 1870, almost totally destroyed by a tornado, during which several of the inhabitants lost their lives.)

We were conveyed from this "City" to the Cave in coaches, the distance being, as before stated, about ten miles, by some estimated at nine, and by others at eleven. (We have recently noticed in the newspapers that, to the great comfort and convenience of visitors, horse-cars have been substituted for the stage-coaches on the route from Cave City to the Cave; but our inquiries, addressed to parties in the neighborhood for a confirmation of this report, have not yet (April 1, 1870) been replied to.) The surface of the country over which this road passes is high, hilly, rocky, and the soil of an apparently poor quality. It is interesting to note the surface appearance along the route, for the reason that, for some distance, this road is supposed to pass directly over a considerable portion of the Cave. At the date of our journey, the latter part of May, this road was in a comparatively good condition; but in the winter and early part of the spring it is said to be almost impassable to travelers. The greater part of the soil is a light colored, sticky clay, with a little sand at intervals. The rocks are composed chiefly of soft white limestone, easily acted upon by chemical and mechanical agencies; hence we find them excavated and jagged, presenting rough, irregular outlines; their outside color is of a dirty, grayish character, owing to exposure to the elements, but the interior is white.

There are small cultivated patches of ground here and there, scarcely deserving the name of farms. The country generally is covered with straggling forests, consisting chiefly of "blackjack," white oak, chestnut, etc. Frequently along the road may be seen small circular depressions in the ground, called "sinks," the surface having fallen in consequence of subterraneous excavation. The whole of the surrounding country appears to be of a cavernous nature; and, if the traveler should be so unfortunate as to possess a timid disposition or large development of caution, he might be apprehensive of a sudden disappearance of the stagecoach into the bowels of the earth.

There are several caves in this vicinity, namely, Proctor's Cave, about three miles in length; White's Cave, Diamond Cave, and the Indian Cave, each of which is about one mile in length.

The Indian Cave opens directly on the stage route; and, as the coaches halt sufficiently long to give visitors an opportunity of examining it, we embraced the occasion for preparing our senses, in this minor cave, for witnessing the stupendous curiosities yet in store for us. An exceedingly loquacious young man acted as our guide. He stated that he discovered the Cave himself, six years previously, and was joint proprietor with his father, who lived near by.

The ingress to this Cave is quite difficult. The descent from the road to the mouth of the Cave is almost perpendicular, and the distance is about one hundred feet. The mouth itself consists of a circular passage about three feet in diameter, and eight feet deep. The descent is made at this point by the aid of rude wooden steps. In answer to an inquiry why greater conveniences for entrance were not provided, we received the unsatisfactory reply that he did not wish to disturb the original appearances of nature.

Upon reaching the foot of the ladder, we found ourselves in an open space, somewhat higher than a man's head, and about ten or fifteen feet wide.

This cave apparently extends in nearly a direct line. We say apparently, for it is impossible for an individual who enters a dark hole under ground, for the first time, to form a correct idea of direction or distance.

The length of this cave, as before remarked, is about one mile. The floor being comparatively smooth, and nearly level, there was but little fatigue attending the exploration.

There is a considerable number of very handsome stalactites and stalagmites to be seen in this Cave, the beauty of which will fully repay visitor for the time thus occupied.

One of the chief curiosities of the Indian Cave is the Pool of Bethesda. It is a fountain of pure, limpid water, about four feet in diameter, and nearly circular in form, and is mantled around with delicate, coral like formation stalagmites, giving it the appearance of a rustic work of art. We partook freely of the water, and found it agreeable to the palate.

Another point of interest is Aline's Dome, said to be named for Miss Aline Du Pont, who, we were told, was the first lady visitor to this Cave. This dome is not of large proportions, but displays more than ordinary beauty, being surrounded by what is known as Elphies's group of stalactites.

There are several other parts of the Cave having fanciful names, possessing more or less interest, but they did not impress us sufficiently to be remembered.

We inquired of our guide why the name "Indian" had been applied to the Cave. He stated that the name was suggested by the fact that, upon his first entrance within the Cave, he discovered several Indian skeletons. Upon manifesting our curiosity to see them, he informed us that, in consequence of the bones having, on different occasions, been sacrilegiously handled by some of the visitors, even to carrying them out and leaving them exposed upon the ground, he considered it his Christian duty to deposit them in a place where they would escape further desecration; he then pointed out to us a deep pit in the Cave, into the invisible depths of which he had thrown them. Visitors may take this explanation as fact or fancy, according to the amount of credulity they possess: in either case, their interest in the Cave need not be lessened.

We reentered the coaches, and, after a ride of about five, miles, reached the Mammoth Cave Hotel, about five o'clock in the afternoon. An exceedingly disagreeable, drizzling rain was falling; and although we were in the southern part of Kentucky, in the latter half of May, we found the atmosphere so chilly as to require the use of fires in our rooms. For the benefit of the Cave visitors of the present day, it is proper to add, in this place, that we have recently (1870) received a communication from Mr. Proctor, of the Cave Hotel, in which lie states that Glasgow Junction, as a stopping place, on the Louisville & Nashville Railroad, for parties visiting the Cave, has various advantages over Cave City: first, it is about three miles nearer the Cave (being but seven miles distant); second, an excellent stage road has been recently made between the points; and, third, immediately upon this route lie the Diamond and Proctor Caves, both of which are exceedingly beautiful and interesting.

With the view of gaining time, some of our party were anxious to enter the Cave on the night of our arrival, thinking that it was a matter of little consequence whether it was day or night on the outside, knowing that perpetual night reigned within. It was soon ascertained, however, that parties were not permitted to enter except at stated hours, at nine and at half past nine o'clock in the morning, according to the route taken. This system was explained as being necessary for the benefit of the guides, and for the proper regulation of the hotel arrangements.

A guide who had been journeying through the Cave all the day of course would not feel willing to continue his travels through the night also. Physical exhaustion, if no other consideration, would render such a procedure impracticable. Our own experience afterward enabled us to appreciate the force of the latter argument. An additional number of guides, undoubtedly, might be kept, but their services would be so rarely required that the proprietors do not feel justified in incurring the extra expense.

This delay gave us an opportunity of taking a survey of the premises.

The Cave Hotel is large and commodious. It is built in the Southern style, with wide verandas, is amply ventilated, and is said to be capable of accommodating between four and five hundred guests at a time. The rooms are of sufficient size, and are very well furnished. The table is really deserving of praise, for it is supplied with the best quality of excellently cooked food, and is accommodatingly attended by experienced negro waiters. A large ballroom is united with the hotel; and is fitted up with all the conveniences required by those who pay court at the shrine of Terpsichore. Connected with the main building, and running at right angles with its front, is a long row of cottages, with a continuous veranda, extending at least three hundred feet.

In speaking upon this point, Bayard Taylor remarks, "The main body of the hotel, with this wing, furnishes at least six hundred feet of portico, forming one of the most delightful promenades imaginable for summer weather."

About one hundred yards beyond the extreme end of the cottages, well shaded by forest trees, may be seen the remains of a tenpin alley building. This went down during the war; and, as the proprietors suffered so severely from the entire loss of business during those four or five gloomy years, it has not yet been rebuilt.

Other marks of dilapidation are also apparent, from the same cause; but as the return tide of visitors begins to flow, with its attendant prosperity, evidences of restoration are visible.

The building and the surrounding grounds are in marked contrast with those seen by the way from Cave City. The visitor is surprised to find in this uncultivated "backwoods" such a large and cheerful looking dwelling and so handsome a lawn. The lawn comprises about two acres of ground, is laid out with gravel walks, and is tastefully ornamented with cedar and other trees.

There are not many summer resorts where an individual or a family can pass a few weeks more pleasantly or more profitably than at the Mammoth Cave Hotel. Here are to be found all the advantages of a first class watering place hotel, with the addition of fine country scenery, and daily opportunities of observing Nature's great subterranean wonder.

In the yard, immediately in front of the main building, stands a very curious looking sandstone rock, about three and a half feet square. One side of the rock has a regular surface which is covered with perforations similar in size and shape (though more widely separated) to the openings in the ordinary cane seated chair, about half an inch in depth, and arranged in regular lines. This rock, we are informed, was excavated near the Cave about twenty years ago. No explanation was offered as to the probable cause of the perforations, but we were left to infer, from their perfect regularity, that they were produced by human agency. Perhaps some rude Indian artist, hundreds of years ago, is entitled to the credit of exciting our curiosity at the present day.

We learned at the hotel that the Mammoth Cave and the Cave Hotel belonged jointly to nine or ten parties, to whom it had been devised by its former proprietor; Dr. Crogan, for a period of ninety nine years. Only about twenty five years of the time have yet expired.

It seems to be regarded by the public as an unfortunate disposition of the property, that so many parties should be concerned in the ownership. Owing to their diverse views, the Cave travel is not so easy nor so agreeable to visitors as it might be made with trifling expenditure. It is said that some of the proprietors are anxious to do one thing, some another, and some nothing. Being unable to agree, nothing is done; and visitors are compelled to undergo much rough and fatiguing travel within the Cave, over loose rocks, etc., which might be rendered, at small expense, comparatively smooth. Hand cars might be introduced and easily made available over more than half the Long Route, stopping as frequently as the curiosity of the visitor might require in making his observations.

Green River, with its towering cliffs, is but a few hundred yards from the hotel. Bayard Taylor, upon first beholding this beautiful river, at the time of his visit to the Cave, sixteen years ago, was struck by the appropriateness of the lines of Bryant, which were applied, strange to say, to another river of the same name:

> "Yet, fair as thou art, thou shunnest to glide, Beautiful
> stream I by the village side, But windest away from haunts
> of men, To silent valley and shaded glen."

It has been conclusively proven, by careful observations, that the rivers of the Cave have a subterraneous communication with Green River.

The entrance of the Cave is about one fourth of a mile from the hotel, and is reached by passing down a wild, rocky ravine through a dense forest, a fitting avenue to the hidden world.

The opening surrounding the mouth of the Cave is irregularly funnel shaped; the walls being steep, and forty or fifty feet in height, and between fifty and one hundred feet across the top of the funnel.

"Trees," says Taylor, "grow around the edges of the pit, almost roofing it with shade; ferns and tangled vines fringe its sides; and a slender stream of water falls from the rocks which arch above the entrance, dropping like a silver veil before the mysterious darkness beyond."

At nearly all seasons a mist or fog may be seen hanging over the mouth of the Cave.

When the external air is warmer than that of the Cave, the mist is produced by the condensation of the moisture of the former by the reduced temperature of the latter. On the contrary, if the temperature of the external atmosphere is lower than that of the Cave, the moisture of the air of the latter is condensed in a similar manner.

When the temperature of the outer air is the same as that of the Cave, no fog or cloud is observable about its mouth.

The entrance of the Mammoth Cave, at an early period of its history, as has already been stated by Mr. Gorin, was situated about half a mile from its present location, constituting what is now called Dickson's Cave. This Cave terminates within a few feet of the mouth of Mammoth Cave, but there is at present no direct communication between the two. The voice of a person at the end of Dickson's Cave can be distinctly heard at the entrance of Mammoth Cave.

The present entrance to Mammoth Cave was formed, and its communication with Dickson's Cave cut off, by the disintegrating action of the water of the spring, which discharges its contents at the mouth of the former, and caused the Cave to fail in at this point, thus establishing a new entrance, and shortening the length of the Cave about half a mile. This is also the theory put forth by Dr. Wright, and there seems to be no reason for questioning its correctness.

Dickson's Cave differs but little in size and appearance from Proctor's Arcade in the Mammoth Cave.

# CHAPTER IV. - ATMOSPHERE OF THE CAVE.

As the circulation of the air, its temperature, purity, etc., in the Cave, are subjects upon which we are frequently interrogated, and which possess great interest to all anticipating a visit, we proceed to give the explanation of these points in the words of Dr. Wright, who thus treats of what he very properly terms the respiration of the cave.

The Mammoth Cave breathes once a year. That is to say, in summer, or when the temperature of the external air is above that of the Cave, the current sets from the latter to the former. In other words, the Cave is the entire summer in making an expiration. On the other hand, when the order is reversed, or the temperature of the outer atmosphere is below fifty nine degrees, the Cave makes an inspiration, or draws in its breath, which it accomplishes during the winter. The respiratory mechanism of the Cave ceases to operate, or to carry out the metaphor, it holds its breath, when the mercury in the thermometer stands at fifty nine degrees in the outer air, which is the average temperature of all parts of the cave, winter and summer. Hence it is frequently observed, in the spring and fall, that there is no motion of the air in either direction at the mouth of the Cave.

On entering the Cave a few hundred yards in summer, when the temperature outside is at or near one hundred degrees, the air rushes out with such force as frequently to extinguish the lamps. Passing into the Cave for about half a mile, however, the motion of the air is barely perceptible at any time, from the fact that the main avenue enlarges so rapidly that it plays the part of a reservoir, where a current of air, from any direction, is speedily neutralized. If the current of air blows from without inward, and is below fifty nine degrees, it does not pass more than a quarter of a mile before it is brought up to that point. Air above the average temperature of the Cave never blows into it.

Thus it will be observed that a change of seasons is unknown in the Mammoth Cave; and day and night, morning and evening, have no existence in this subterranean world. In fact, there is an eternal sameness here, which is without a parallel.

In many parts of the Cave, time itself is not an element of change; for where there is no variation of the temperature, no water, and no light, the three great forces of geological transformation cease to operate.

The atmosphere of the Cave, contrary to what might be generally supposed, is remarkably pure and wholesome.

The proportions of oxygen and nitrogen bear the same relation to each other in the Mammoth Cave that they do in the external air. The proportion of carbonic acid gas is less than that observed in the atmosphere in the surrounding country, upon an average of many observations. This noxious gas is one of the necessary constituents of vegetable existence; and, as there is no vegetable life within the Cave, its comparative absence is a natural inference.

In the dry parts of the Cave the proportion of carbonic acid is said to be about 2 to 10,000 of air; in the vicinity of the rivers, something less. Not a trace of ammonia can be detected in those parts of the Cave not commonly visited.

The amount of the vapor of water varies. Thus, in those avenues at a great distance from the rivers, upon the walls and floors of which there is a deposit of the nitrate of lime, the air is almost entirely destitute of moisture, from the hygroscopic properties of that salt; and animal matter mummifies instead of undergoing putrefactive decomposition. For the same reason, no matter what state of division the disintegrated rock may attain, dust never rises. In portions of the Cave remote from the localities in which the bats hibernate, no organic matter can be recognized by the most delicate tests. Not a trace of ozone can be detected by the most sensitive reagents.

From what has been stated, it will be observed that the atmosphere of the Mammoth Cave is freer from those substances which are calculated to exert a depressing and septic influence on the animal economy than that of any other locality on the globe. This great difference is observed by every one on leaving the Cave, after having remained in it for a number of hours.

In such instances, the impurity of the external air is almost insufferably offensive to the sense of smell, and the romance of a "pure country air" is forever dissipated.

The only instance that history (possibly romance) records, so far as is known to the writer, in which these disagreeable effects of the ordinary atmosphere were markedly produced, was in the case of the unhappy

Caspar Hauser, who was confined in a subterranean dungeon at Nuremberg from infancy to adult age. When he was finally brought upon the surface of the earth, his life was rendered miserable by the insufferable odors that constantly impressed his olfactory nerves. The smell of flowers, that to others were sweet, was so intensified in his case as to be exceedingly disagreeable. He was unable to pass a graveyard, where others could detect no odor whatever, without fainting from the painful impression received through the sense of smell. This shows that, to appreciate "country air," our senses must be adapted to it by constant contact.

# CHAPTER V. - THE FORMATION OF THE CAVE.

Before entering the Cave, it will be proper for us to consider the agencies concerned in its formation. These may be divided into chemical and mechanical. We strictly follow the words of Dr. Wright in these explanations, knowing that his education in this particular, and his opportunities for observation, eminently qualify him for giving correct views on the subject.

Of the chemical agencies, which were undoubtedly the most remarkable and important, he says, "There can be no doubt but that the solvent action of water holding carbonic acid in solution was the primary agency concerned in the formation of the Cave. Thus the limestone, or carbonate of lime, which constitutes the strata of rock through which the Cave runs, is not soluble in water until it combines with an additional proportion of carbonic acid, by which it is transformed into the bicarbonate of lime. In this way the process of excavation was conducted, until communications were established with running water, by which the mechanical agency of that fluid was made to assist the chemical. The little niches and recesses which are observed in various parts of the Cave, and which seem to have been chiseled out and polished by artificial means, were formed in this manner; for when these points are closely examined, a crevice will be observed at the top or back of them, through which water issued at the time of their formation, but which has been partially closed by crystals of carbonate of lime or gypsum. At the time these niches were forming, water flowed through the avenues in which they are found. Examples of the action we have been describing may be seen in Sparks Avenue, leading to the Mammoth Dome.

"The grooves which are observed in rock over which water is or has been flowing are also formed by the solvent action of water containing carbonic acid; for in all such instances the water has no solid matter in suspension. Examples of this kind of action may be seen in operation in Mammoth and Gorin's Domes; and evidences of its former action may be observed in Lucy's Dome. What are termed the pigeon holes in the Main Cave are cut out of the solid rock in the same manner.

"Another agency which contributes in part to change the appearance of the Cave is the efflorescence of the sulphate of soda, or glauber-salts, and the crystallization of sulphate of lime, or plaster of' Paris.

"The sulphate of lime, which is known under the names of gypsum, plaster of Paris, selenite, alabaster, etc., exerts a much greater influence in disintegrating rock than the sulphate of soda. The avenues in which gypsum occurs are perfectly dry, differing in that respect from those that contain stalactites. When rosettes of alabaster are formed in the same avenue with stalactites, the water which formed the latter has for ages ceased to flow, or they are situated far apart, as the former cannot form in a damp atmosphere. The force exerted by gypsum in the act of crystallizing is about equal to that of water when freezing, and when it crystallizes between ledges of rock, they are fractured in every direction, as instanced in Pensacola Avenue and Rhoda's Arcade.

"The formation of nitre is due, in part, to the decomposition of bats and other animals; but it must not be forgotten that limestone rocks are never entirely destitute of nitrifiable matter. The nitric acid which enters into its composition may, in some measure, be derived from the atmosphere. The kind of nitre that is found in the Cave is the nitrate of lime, which, when reacted upon by the carbonate of potash, is transformed into the nitrate of potash, or common saltpetre. This was the course pursued by the saltpetre miners when that substance was manufactured in the Cave in 1812-14. The nitrate of lime is found in the dryer parts of the Cave, but is not discoverable till the earth which contains it is lixiviated.

"The mechanical agencies concerned in the excavation of the Mammoth Cave are trifling when compared with the chemical. They are instanced in the transportation of gravel, sand, and clay from one part of the Cave to another, and in the abraded appearances presented by the rock composing certain avenues. Thus, it is possible to tell the direction in which the water ran in most of the avenues, and the rapidity of its motion, by observing the points at which gravel, sand, and clay are deposited, and the order in which they come. For example, the points at which gravel is deposited indicate a rapid current; where sand is found, the movement was slower; and where clay occurs, the water was almost or quite stationary.

"At one time the water rushed with great force through Fat Man's Misery, for in Great Relief, which is just beyond, washed gravel occurs; still farther, sand is found, which is succeeded by clay: showing that the current was in the direction of Echo River. Before the mechanical agency could have exerted any appreciable influence, the chemical must have been in operation for thousands of ages.

"The loose rocks that are scattered on the floors of many of the avenues have fallen from the walls and ceiling, but in many instances the points from which they were detached are indistinct, from the fact that the rugged surface from which they have fallen is either smoothed by the action of water, or covered by crystals of the carbonate or sulphate of lime.

"In those parts of the Cave where no rocks have fallen, the floor presents the appearance of the bed of a river, and is covered with gravel, sand, or clay, according to the rapidity of the flow of the water at the time of the deposit.

"Visitors need feel no apprehension or alarm in reference to falling rocks, for none have fallen since the discovery of the Cave."

It may be well in this place to refer to the interesting relation subsisting between Mammoth Cave and Green River. There can be no doubt that Green River has cut out the bed or channel through which it runs; for on ascending its banks on either side for a distance of not less than three hundred feet, a plain is reached, which is not succeeded by a valley; establishing conclusively that it has worn its bed to its present level by the mechanical and chemical agency of water, and that the avenues of the Cave were cut through with nearly equal pace, those near the surface of the earth being formed first, and the others in regular order from above downward; the avenues through which Echo and Roaring Rivers run being the lowest and last formed. Both of these rivers are on a level with Green River, with which there is, as before stated, a subterraneous communication. As Green River continues to deepen the valley through which it passes, the avenues of the Cave will continue to descend, until the springs which supply Echo and Roaring Rivers cease to flow, when the avenues through which they run will become as dry as Marion's Avenue, which, at an early period in the history of the Cave, contained the most beautiful subterranean river in the world.

With these preliminary details, which we consider essential to a proper understanding or an intelligent appreciation of the curiosities of the Cave, we will proceed to conduct the reader within its portals.

## CHAPTER VI. - THE LONG ROUTE.

ON the morning succeeding our arrival at the Cave Hotel, our party, consisting of fifteen persons, seven of the number being ladies, fully equipped in Cave costume, left the house at nine o'clock precisely, to explore what is known as the "Long Route," which terminates at the Maelstrom. For the information of the uninitiated, we will explain that the costumes referred to are kept at the hotel for the use of the visitors. It is necessary for its greater convenience in threading narrow passages, and for the equally important object of preserving from damage more expensive clothing. Ladies are provided with short dresses of stout material, generally of fancy and picturesque colors, without the addition of crinoline. Gentlemen have short woolen jackets, caps, and "overalls." Many jocular remarks are usually made by parties, thus oddly attired, at each other's expense.

Our party being large, and one or two of the number being in somewhat feeble health from recent indisposition, we deemed it prudent to employ two guides to accompany us, so that one of them might return from any point with such parties as might become either unable or unwilling to proceed, while the other could conduct those who wished to continue the journey. This proved to be a wise precaution, as one or two of the ladies became too much fatigued to be able to complete more than about two-thirds of the route.

The present guides at the Cave are white men; and the chief one in charge of our party, Mr. Charles Merideth, is a man of considerable intelligence, is well versed in all matters pertaining to the Cave, and, in common with the other guides, is fully qualified for the performance of the duties of his important office.

The guidance through the Cave was formerly under the charge of colored men. Several of them, Stephen, Alfred, and Mat, attained great celebrity in this capacity; and all former visitors remember these names as a part of their Cave experiences.

Stephen, who was particularly famed for his qualifications in this respect, as has been seen in Mr. Gorin's remarks, after a long and honorable career in exhibiting and explaining the curiosities of the Cave, with which his name has become identified, to thousands of delighted visitors, departed this life about eleven years ago.

Alfred is also dead. "Old Mat," as he is familiarly called, who has trodden the dark and mysterious paths of the Cave for more than thirty years, still lives, and may be seen about the hotel, but is no longer on duty, yet he thinks he is quite as capable of exhibiting the Cave now as he ever was, and believes that he possesses more knowledge regarding it than any one else.

From the hotel we passed down the deep ravine through the native forest, before mentioned, along the rugged pathway. The precipitous and rocky character of the path, however, was not particularly observed until our return at night. We then began to wonder if some freak of nature had not occurred in our absence to cause the picturesque and rather easy graded path of the morning to present a nearly perpendicular front, every small stone that lay in the way to attain the proportions of an insurmountable rock, and the fourth of a mile that we had passed so easily and so pleasantly in the morning to be lengthened out to at least three times that distance. Then we were ready to exclaim, "0, for a horse!"

Upon reaching the entrance, which we do by descending the steep bank leading to it by means of rough stone steps, the guides immediately proceed to light the lamps, which are kept within the mouth of the Cave for the use of visitors.

Proceeding a few steps, each with lamp in hand, we plunged into almost total darkness, our aids to sight appearing to afford but little light to our unaccustomed eyes. We were ready to despair of ever getting a view of the beauties of the Cave with such limited means of illumination. But in a few moments, our pupils having had time to expand, and adapt themselves to the sudden change from light to darkness, we were gratified to discover that we could obtain a very satisfactory view of the dark interior.

Upon entering the Cave for the first time, we feel the force of the words of Dante:

"Who enters here leaves hope behind." This is literally true, but not however in the terrible sense implied by the poet. We not only leave hope, but we leave care and sorrow and all the feelings that make up the sum of our mundane existence, in the world behind us. We really enter a new phase of life. We forget, for a time, the life we have lived before. Here we find no objects of comparison, nothing to remind us of our pre-existence. It is worth a visit to the Cave to experience these new and extraordinary sensations.

We first enter a small archway at the mouth of the Cave, called the Narrows. The sides are walled up with rock, which the saltpetre manufacturers removed from the floor at this point to allow of easy ingress.

After leaving the Narrows, the ceiling of which is about seven feet high, and which does not possess any special interest, the Rotunda is entered.

The Rotunda is said to be situated immediately under the dining room of the Cave Hotel. The ceiling of the Rotunda is about one hundred feet high, and its greatest diameter is one hundred and seventy five feet.

The floor is strewn with the remains of vats, water pipes, and other materials used by the saltpetre miners in 1812. The wood of which they are made is in a remarkable state of preservation.

To the right of the Rotunda, Audubon's Avenue leads of for about half a mile, to a collection of stalactites. During the winter, millions of bats hibernate in this avenue. At the entrance of Audubon's Avenue several small cottages, which were built for the residence of persons afflicted with consumption, are still to be seen.

On leaving the Rotunda and passing the huge overhanging cliffs to the left, which are called the Kentucky River Cliffs, from their close resemblance to the cliffs of that river, the Methodist Church is entered.

This apartment is eighty feet in diameter, by about forty in height. Here, we are told, from the gallery or pulpit, which consists of a ledge of rocks twenty five feet in height, the Gospel was expounded more than fifty years ago. The logs used as benches occupy the same position which they did when first placed in the church.

It is customary for visitors to leave their shawls or overcoats, if required outside, at this point, there being no variation of temperature beyond.

Next in order is "Wandering Willie's Spring," a beautifully fluted niche in the left hand wall, caused by the continual attrition of water trickling down into a basin below. This spring is said to have derived its name from an eccentric young country violinist, who, in the spirit of romance, assumed the name of Wandering Willie. He became separated from his companions while within the Cave, had his lamp extinguished, and was found lying asleep beside the spring. This spring is about half a mile from the entrance of the Cave.

We pass the Gothic Galleries, which lead to Gothic Avenue, of which we shall have occasion to speak hereafter, and the Grand Arch is entered, which leads to the Giant's Coffin. This arch is about fifty feet high and sixty wide.

The Standing Rocks are found to the left of the path; they are many tons in weight, and have evidently fallen from above, standing with the base upwards, extending eight or ten feet above the floor. They maintain their upright position from the fact that the earth was penetrated in the fall while in a soft state. The avenue, however, has been perfectly dry since its first discovery.

A short distance beyond, on the right, the guide bids us stop, and asks what we see before us. We hold up our lamps, and all cry out simultaneously, in an awe struck tone, "A coffin!" We are then informed that we behold the Giant's Coffin. This immense sarcophagus is a huge rock, forty feet long, twenty wide, and eight in depth, and, at the point from which it is viewed, presents a striking resemblance to a coffin. It has been detached from the side of the avenue against which it rests.

On the ceiling, a little to the left of the Giant's Coffin, and looking into the Deserted Chamber, is the figure of an anteater. It is composed of the efflorescence of black gypsum, and rests upon a background of white limestone. Bayard Taylor, whose extensive travels enable him to speak authoritatively, says that the resemblance of the figure to the animal after which it is named is very perfect.

A short distance beyond the Giant's Coffin, in the Main Cave, after passing what is called the Acute Angle, a group of figures is observed on the ceiling, termed the Giant, Wife, and Child. These, figures are in a sitting posture, and the Giant appears to be in the act of passing the Child to the Giantess. They are also composed of black gypsum, which rests on a white background.

Still farther on, the figure of a colossal mammoth may be seen on the ceiling.

From the Giant's Coffin to the mouth of the Cave, wheel tracks, and. the impressions of the feet of the oxen used to cart the saltpetre, made over fifty years ago, may be distinctly seen. The earth at the time that these impressions were made, was in a moist condition, having recently undergone the process of lixiviation in the manufacture of saltpetre, and, upon drying, attained an almost stony solidity. These tracks are on the immediate route of travel, and have been walked over by thousands of visitors during a period of sixty years. Yet the cleft foot of the ox, and the regular indentations of the cart wheel, can be plainly distinguished in the petrified earth. At one point we were shown where the oxen were fed; and, by the aid of a stick, we succeeded in digging out of the dry earth two or three impacted corncobs, in a good state of preservation; and we are perfectly satisfied that they had not been placed there for purposes of deception, as has been suggested by some parties.

We were puzzled, at first, to understand how the oxen and carts could be got into the Cave, the descent to the entrance being so precipitous and the mouth so contracted. The guide suggested that the oxen were introduced separately, and the carts in piecemeal.

On the route from the Acute Angle to the Star Chamber, several stone cottages, formerly inhabited by the invalids already mentioned, are still standing, gloomy monuments of their departed occupants. One of these cottages is used as a card room, where hundreds of private and business cards may be found.

We are now upon what is known as the Long Route, and we leave the Main Cave at the foot of the Giant's Coffin.

One by one we pass into a crevice behind the Coffin, at the bottom whereof yawns a narrow hole. Half stooping, half crawling, we descend through an irregular, contracted passage to a basement hall called the Deserted Chamber.

The Deserted Chamber is a gloomy, abandoned looking hall, and is fully entitled to the name given it. This is the point at which the water left the Main Cave to reach Echo River, after it had ceased to flow out of the mouth of the former into Green River. In other respects it is not of particular interest.

The two illustrations which accompany this part of our text "Entrance to the Long Route," and the "Deserted Chamber" give the reader a very correct idea of the singularly wild and extraordinary surroundings in this part of the Cave. The entrance to the Long Route is effected, as has been already stated, through the narrow passage around the far end of the Giant's Coffin. The guide is seen just entering the contracted avenue. The next view represents the guide as having accomplished the passage upon which we saw him entering, and as having reached the dreary looking "Deserted Chamber." He carries upon his arm the basket of provisions for dinner. This chamber is about one hundred feet in length, but the ceiling, as may be seen, is quite low.

(Our lithographic plates are copied from *photographs,* to the perfect accuracy of which we can testify. Forty two stereoscopic views, taken within and about the Cave, have been published. The interior views were obtained by the aid of the magnesium light, the most intense artificial light that has yet been produced. This set of views, which we recommend to the attention of our readers, constitutes a novel and most triumphant application of the photographic art, and materially aids in the comprehension of our language as we treat upon these unique curiosities. They are published by Messrs. Anthony & Co., of New York.)

*Entrance to Long Route*

*Deserted Chamber*

43

An apartment known as the Wooden Bowl Cave is next entered. It derives its name from the tradition that a wooden bowl, such as was formerly used by the Indians, was found in it when it was first discovered. The chamber itself is shaped like an inverted bowl, which fact may have suggested the name.

It is said that the Indians formerly explored the Cave with long reeds, filled with deer's fat, to light them along.

Black Snake Avenue, which enters the Main Cave near the stone cottages, communicates with Wooden Bowl Cave. It receives its name from its serpentine course and black walls. It is rarely shown to visitors, as it possesses but few objects of interest.

We next pass a steep declivity and a flight of steps, called the Steps of Time, and enter Martha's Palace. The Palace is about forty feet in height, and sixty in diameter. It is not particularly attractive, and it appears singular that it should have been accorded so grand a name. A short distance beyond Martha's Palace is a spring of clear, potable water, which visitors generally take advantage of to quench their thirst, as there is a considerable distance, in some parts of the Cave, between the fountains of good drinking water.

The Side Saddle Pit, over which rests a dome sixty feet in height, is reached by passing through what is called the Arched Way; the walls, floor, and ceiling of which bear evidence that it was once the channel of running water. This Pit is ninety feet deep, and at its widest part about twenty feet across.

*Bottomless Pit and Bridge of Sigh.*

About twenty feet to the left of the Side Saddle Pit is situated Minerva's Dome. It is fifty feet in height, and ten in width. It is a miniature representation of Gorin's Dome, hereafter to be noticed. The Dome and Pit have been cut out of the solid rock by the solvent action of water containing carbonic acid in solution. They are still enlarging.

The aperture leading to the Pit presents the outlines of a lady's saddle; hence the name.

We next arrive at the brink of the Bottomless Pit. The very name causes us to shrink with terror; but we are presently reassured by finding it to be a misnomer. The Pit, which doubtless appeared bottomless to the first discoverers, if we credit Mr. Horace Martin, has since been found to be but one hundred and seventy five feet in depth.

The Bottomless Pit was formerly the limit of excursions in this direction. It was not until the year 1838, we are informed, that any traveler ever passed beyond this frightful chasm. In that year the Pit, as has been stated in Mr. Gorin's letter, was spanned by a substantial wooden bridge known as the "Bridge of Sighs;" and then was discovered the most beautiful and interesting portion of the Cave.

Shelby's Dome, which is sixty feet in height, rests directly over the Bottomless Pit. The Pit and Dome have been formed, and are still enlarging, by the same causes that excavated the Side-Saddle Pit.

Immediately beyond the Bottomless Pit a room is entered, called the Revelers' Hall, which is about twenty feet in height, and forty in diameter.

Here it is the custom of visitors to rest for a short time and discuss the terrors of the Pit. This's generally followed by bringing forth the potables, when the safety and health of all parties are duly toasted. So says Dr. Wright; and so will every visitor say when he observes the immense quantity of broken and unbroken bottles strewn about the floor of this wild looking Hall.

After passing through a low archway, about four feet in height, very properly termed the Valley of Humility, the ceiling of which is smooth and white and appears as though it had been plastered, the Scotchman's Trap is entered. The Trap is a circular opening, through which it is necessary to pass by descending a flight of steps. It is about five feet in diameter, over which is suspended a huge rock, like a dead-fall, by an apparently slight support, which, if it were to fall, would completely close the avenue leading to Echo River. If, however, this opening should become closed, we will state, for the comfort of the timid, that there are three ways by which an escape might be effected. Thus: there is an avenue beyond it, which enters the bottom of the Bottomless Pit, from which a person might be drawn up by means of ropes; another avenue of escape would be by Bunyan's Way, which leads into Pensacola Avenue; and a third, by Sparks' Avenue and Mammoth Dome. The accompanying figure shows this Trap, with the guide standing at the head of the steps.

*View from Bridge of Sighs*

*Scotchman's Trap*

A short distance beyond the Scotchman's Trap, in what is termed the Lower Branch, there is found a curiously shaped rock, named the Shanghai Chicken, from its fancied resemblance to that unsightly fowl.

The next curiosity of note that is reached in our progress is one possessing great interest to men, and to women also, who are blessed with a respectable physical development. This place of attraction has been accorded the euphonious name of "Fat Man's Misery:"

Fat Man's Misery is a narrow, tortuous avenue, fifty yards in length, which has been cut out of the solid rock by the mechanical action of the water. The lower part of the avenue varies in width from eighteen inches to three feet; and the upper part, that is, from the height of a man's chest to the head, from four to ten feet. In height it varies from four to eight feet, the greater part of the distance averaging but four feet, thus requiring the passenger to assume a stooping position, which is exceedingly painful to the back.

Contrary to the general impression, says Dr. Wright, there never was a man too large to pass through Fat Man's Misery. This is an error. We have known more than one individual, weighing over four hundred pounds, who could not possibly have effected the passage. Bayard Taylor says that the weight of the largest man who ever accomplished this narrow way was two hundred and sixty pounds, and he thinks that it would be impossible for a man of greater weight to see the sights beyond.

*Bacon Chamber*

A hall of novel appearance, very appropriately denominated Great Relief, after the experience of bended backs and compressed sides in the passage of Fat Man's Misery, is next entered. This hall varies in width from forty to sixty feet, and in height from five to twenty feet. From the ceiling project immense nodules of ferruginous limestone.

On the floor of Great Relief, the direction of the current of water that filled these avenues can be traced. Thus, at the side next Fat Man's Misery it is strewn with gravel, near the center sand occurs, and still farther on mud is deposited, demonstrating the fact that it flowed into Echo River.

The avenue termed Bunyan's Way passes directly over Great Relief; and enters a short distance from Fat Man's Misery, by which communication is established with Pensacola Avenue.

The portion of the avenue in advance, which extends from Great Relief to the River Styx, is called River Hall. It varies in width from forty to sixty feet.

The Bacon Chamber is situated to the right of River Hall. This chamber is decidedly curious, and the name singularly appropriate. Here may be seen a fine collection of limestone hams and shoulders suspended from the ceiling, as in a smokehouse. They were formed by the solvent action of water charged with carbonic acid, at the time when the lower portion of them rested against a stratum of rock which has since been detached.

The avenue which leads to the Mammoth Dome and Sparks' Avenue takes its origin in the Bacon Chamber.

About forty feet below the terrace which leads to the Natural Bridge is a body of water, fifteen feet deep, twenty wide, and fifty feet in length, termed the Dead Sea. It is quite as gloomy, we are told, as its celebrated namesake. Mr. Martin says, "The name so awful and so referable to awful events cannot be better illustrated than here. There is a terrible grandeur in the place. Long after you have left it, the mind's eye continues cognizant of its many sights, the ear of its many sounds. The memory holds them, and they ever haunt the dreams of night."

When this part of the Cave was first discovered, the Dead Sea was passed on the terrace over its left bank; this passage, however, was attended with great danger.

By a curious anomaly, our teachings in heathen mythology are reversed in the Mammoth Cave. Here we pass the Bottomless Pit before reaching the River Styx, instead of ferrying over the latter on our way to the former!

The "Visitor" (whose work was published by Morton & Griswold, Louisville) remarks, "He who could paint the infinite variety of creation can alone give an adequate idea of this marvelous region. As you pass along, you hear the roar of invisible waterfalls; and at the foot of the slope the River Styx lies before you, deep and black, overarched with rock. The first glimpse of it brings to mind the descent of Ulysses into hell,

> "Where the dark rock o'erhangs the infernal lake,
>
> And mingling streams eternal murmurs make."

The River Styx is one hundred and fifty yards long, from fifteen to forty in width, and in depth varies from thirty to forty feet. It has a subterranean communication with other rivers of the Cave, and, when

Green River rises to a considerable height, has an open communication with all of them.

The Natural Bridge spans the River Styx, and is about thirty feet above it. When the farther bank of the River Styx is illuminated with a Bengal light, the view from the Natural Bridge is awfully sublime.

Our attention is next drawn to a silent, peaceful looking body of water, called Lake Lethe. This lake is one hundred and fifty yards long, from ten to forty feet wide, and in depth varies from three to thirty feet. The ceiling of the avenue at this point is ninety feet above the surface of the lake. Lake Lethe extends in the direction of the avenue, the floor of which is covered by it.

The lake is crossed in boats. On the occasion of our visit the boat was not sufficiently large to carry all of our party at one time; it was therefore necessary that a number of us should remain for the second trip. We sat down upon the dark shore and watched the boat glide slowly away. The novel scene was peculiarly adapted to the production of a lasting impression upon the imagination of the beholder, the boat moving slowly and noiselessly over the water, carrying its phantom like freight, dressed in their fanciful costumes, the dim lamps throwing fitful flashes of light and shadow on the rippled surface, and through the darkness to the high ceiling above; then, as we silently gazed, with unutterable thoughts, the boat and its specter like voyagers passed entirely from our view around a projecting angle of rock; darkness reigned upon the face of the waters, as in primeval chaos; a long breath was taken, and some abortive efforts were made to express our feelings. After a brief interval of darkness, the Charon of this stream, with his solitary lamp in the prow of his rude boat, reappears in the distance, returning for those left behind. The feelings inspired by this scene, we say, were of a character that can never be forgotten, and such, perhaps, as could be experienced under no other circumstances; for no counterpart of the surroundings are known to exist. Being fatigued and thirsty, on our return from far beyond, we drank of the waters of Lethe, without, however, forgetting our troubles, sore feet and weak knees!

Upon disembarking on the opposite shore of Lake Lethe, we enter Great Walk, which extends from the lake to Echo River, a distance of five hundred yards.

The ceiling is forty feet high, and the rocks which compose it present a striking resemblance to cumulus clouds. They are composed of white limestone. The floor is covered with yellow sand.

It requires a rise of only five feet of water in Echo River to overflow Great Walk; and that depth is sufficient to allow boats to float between the lake and the river. There arc times, we are informed, when Great Walk is filled with water from the floor to the ceiling. Extraordinary as the statement may appear, it is not an uncommon occurrence for the water to rise to the height of sixty feet in Lake Lethe, at which times the iron railing on the terrace above the Dead Sea is entirely submerged. This great rise of water is produced by freshets in Green River.

# CHAPTER VII. - ECHO RIVER.

We next arrive at the banks of Echo River.

"Darkly thou glidest onward,

Thou deep and hidden wave!

The laughing sunshine bath not look'd

Into thy secret Cave.

"Thy current makes no music,

A hollow sound we hear,

A muffled voice of mystery,

And know that thou art near.

"No bright line of verdure

Follows thy lonely way,

No fairy moss or lily's cup

Is freshened by thy play."

Connected with this river are perhaps, some of the most delightful of the multitude of impressions that we receive in the Cave. There are sights more gorgeous, more awful, more sublime, but nowhere are the senses of sight and sound so beautifully and so charmingly brought into unison. In point of sublimity, impressing the senses through the sight alone, the Star Chamber, in the Main Cave, excels it; but all who are capable of being agreeably affected by the "concord of sweet sounds" will recall the voyage over Echo River as the most charming reminiscence connected with their visit to the Mammoth Cave. It is the fairy river that wafts upon its bosom the wandering traveler to the mystic regions beyond.

Echo River extends from Great Walk to the commencement of Silliman's Avenue, a distance of three quarters of a mile.

The avenue at the entrance of Echo River, under ordinary circumstances, is about three feet in height, which, as can be easily imagined, is rather a contracted space for a boat with its human, freight to pass beneath. A large flatboat is kept here, which we found large enough to carry the entire number of our party at a single trip. Considerable stooping was necessary to pass under this low arch on our outward bound voyage; but before our return the river had risen several inches, so that it was necessary to get down on the hands and knees, and even lower, in order to pass the arch. The unpleasantness of the situation may be inferred, when it is stated that the boat, in consequence of being frequently submerged by the rise of the river, is always wet and muddy in the interior. In effecting our exit from this narrow passage on our return voyage, some ludicrous incidents occurred, owing to the necessary sacrifice of grace and decorum on the part of the ladies, as well as on that of the gentlemen; some of the party barely escaping being crushed by the unexpectedly sudden descent of heavier individuals.

If a disinterested observer could have witnessed the scene at this point, the entire party in every possible awkward position, stooping low, lying down, some lustily crying out that they were being crushed by somebody, some laughing, and some complaining that the ceiling had damaged their heads, we repeat, if a disinterested observer had been present, the scene would have appeared to him as ludicrous in the extreme.

Fortunately, however, for the comfort of visitors, this low ceiling does not extend more than fifteen or twenty feet from the entrance; beyond that distance the average height is about fifteen feet. At some points the river is two hundred feet wide. In depth it varies from ten to thirty feet. The ceiling is of an arched form, and is composed of smooth, solid rock, more closely resembling a work of art than of Nature.

From what has been said of the narrow opening at the starting point on the river, it may be inferred that a slight increase of water would render ingress impossible. There is a means of escape, however, should any one be caught beyond, by a small side avenue, called Purgatory, which commences at the end of Great Walk, and terminates in the avenue of Echo River, about a quarter of a mile from the landing in Silliman's Avenue. A rise of eighteen feet of water, however, fills the avenue of Purgatory, and cuts off all communication with the outer world.

When there is no rise in Green River for several weeks, the water in Echo River becomes remarkably transparent, so much so, in fact, that rocks can be seen ten or twenty feet below the surface, and the additional. novelty is given to the voyage of the sensation that the boat is gliding through the air. The connection between Echo and Green Rivers is doubtless near the commencement of Silliman's Avenue. When Green River is rising, Echo River runs in the direction of Great Walk; when it is falling, the current sets in the opposite direction. When Green River is neither rising nor falling, the water of Echo River runs slowly in the direction of Silliman's Avenue, and is supplied from springs in the Cave. At such times its temperature is fifty nine degrees, the same as the uniform temperature of the atmosphere of the Cave. When the water of Green River flows into Echo River at a temperature higher than that of the Cave, a fog is produced, which in point of density, it is said, is not inferior to that off the banks of Newfoundland. Inexperienced persons have been lost in the fog on Echo River.

At the time of our voyage across this river there was no fog, and the water, though not transparent, was beautifully clear. After proceeding a short distance, the guide, who stood in the bow of the boat, silently propelling it by means of his hands, when within reach, and, at other times, by a staff applied to the ceiling and side walls, struck up at short intervals, a plaintive note of song. From the far distance, as from another world, we had almost said from the spirit world, came answering melodies, as though a thousand tongues, attuned to different chords, had taken up the refrain, repeating it again and again, fainter and fainter, whilst we unconsciously strained our ears and stayed our breathing to catch the last dying tone. Here, one, without effort of imagination, might easily conceive that he was really passing over the "dark river," and within the sound of the choristers that stand upon the celestial shores to welcome him onward!

Lord Byron has beautifully described the echo of thunder among the mountains:

"Far along,

From peak to peak, the rattling crags among,

Leaps the live thunder! Not from one lone cloud,

But every mountain now hath found a tongue,

And Jura answers, through her misty shroud,

Back to the joyous Alps, who call to her aloud!"

This picture lifts us to the sublime and inspires us with awe; but on Echo River all is calmness and peace, harmony and love, we forget the world behind us, we forget our pre-existence, we realize our ideal of an approach to spiritual life.

Two or three of the gentlemen of our party, in a spirit of adventure, made their passage through the rugged avenue called Purgatory. They described the trip as one of exceeding difficulty. After terminating their purgatorial experience and arriving at the end of the avenue on Echo River, which they accomplished several minutes in advance of the boat party, one of the gentlemen fired a pistol. So remarkable was the effect that it sounded to our astonished ears like the explosion of heavy artillery, reverberating for a surprising length of time.

# CHAPTER VIII. - THE EYELESS FISHES OF THE CAVE.

As an interlude to the descriptive narrative of the scenery observed in our journey through the Cave, we will stop here to note the existence of animal life, which is remarkable for maintaining vitality under circumstances so unfavorable for normal development. There are to be found in the Cave eyeless fish, eyeless crawfish, lizards, frogs, crickets, rats, bats, etc., all, except the two first named, being possessed of more or less development of the visual organs.

In Echo River we find the eyeless fish and the eyeless crawfish. These specimens of the fish tribe have been looked upon by all classes of persons, ever since the first published notice of their existence, as remarkable curiosities. They illustrate, however, a fixed rule in the great laws of Nature. The presiding Deity never supplies any living creature with superfluous organs; and if organs already exist which future circumstances render useless, they are eventually obliterated. In the never ceasing darkness of the Cave, eyes are unnecessary organs to the fish that live in its waters. We have every reason to believe that these fish were originally possessed of eyes; but after their introduction into the Cave, and perhaps centuries of existence there, these useless organs gradually, through many generations, lost their original character, and finally disappeared; only a trace of the orbit remaining. ("Their exclusion from the solar beam is well known to produce organic alterations in the visual organs of animals, such as atrophy of the optic nerve, or those portions of the brain (the *corpora quadrigemina*) more immediately associated with the sight. It is supposed that the blindness observed among fish found in the dark caves of the Tyrol and Kentucky arises from the arrest in the development of the eyes as the result of a constant deprivation of light." *Light: its Influence on Life and Health.* By Forbes Winslow, M.D., etc., American ed. New York: Moorhead, Simpson & Bond, 1868, p. 13.)

Some months after our visit to the Cave our attention was drawn to a newspaper article, from an anonymous writer, which we believe originally appeared in the Chicago "Tribune," August 18, 1867. The writer contended that the permanent inhabitants of the Cave were not only blind, but deaf also. The original letter in the "Tribune" was entitled "Important Scientific Observations," etc., made in the Mammoth Cave. Feeling interested in everything connected with the Cave, particularly the scientific observations, we read the letter with more than ordinary attention, but were disappointed to discover that mere conjectures of a sensational character were presented to the uninformed public as the result of scientific investigations. The writer begins by saying, "Will you permit me through the columns of your paper to invite attention to some very remarkable natural facts, communicated to me by Dr. (naming a physician), of this city, which came under his observation during a visit of scientific research to that geological freak of nature, the Mammoth Cave? They seem to be worthy of record, but, as the doctor modestly intimated, may have been the subject of observation by others as well as himself, although perhaps not possessing the same degree of interest. Keenly alive to everything, however remotely connected with his favorite profession, the doctor, it seems, was perfectly astonished at the fixed and chronic state of blindness and deafness in which he found the permanent inhabitants of the Cave. These beings, it appears, are not only without eyes, or even the trace of an orbit, but, so far as could be ascertained by careful and indefatigable investigation, evidently destitute of the sense of hearing." This writer bases his theory upon the assumption that there is no sound in the Cave to produce vibrations upon the auditory nerve, forgetting that the animals, the rats particularly, cause sounds by their own voices and movements. He does not confine his remarks, as to the deficiency of sight and hearing, to the fishes, but includes all "the permanent inhabitants of the Cave."

A portion of the said letter was afterward copied in some of the medical journals, and, among others, in the "Medical and Surgical Reporter," of Philadelphia, vol. xvii. p. 479 (Nov. 30, 1867). We took occasion in a subsequent number of the same journal to express our dissent from the promulgation of such unsupported assertions, and called upon the author for the experiments which were said to have been instituted for determining the absence of the organ of hearing in the inhabitants of the Cave. He replied in an evasive and somewhat surly manner, without giving any experiments or arguments to sustain his theory.

We will first describe the general characteristics of the fish, and afterward recur to the point above referred to.

The fish are of a peculiar species, and are of a class known as viviparous, which give birth to their young alive, and do not deposit eggs after the manner of most other fish. They have rudiments of eyes, but no optic nerve, and are therefore incapable of being affected by any degree of light. We are indebted to Dr. Wright, who is perfectly familiar with the facts, for this statement.

The eyeless crawfish give birth to their young in the same manner as those provided with eyes. Both the fish and the crawfish are of a color almost white.

Ordinary fish and crawfish are sometimes washed into the Cave from Green River. Frogs, also, are occasionally washed into Echo River, and, at times, may be heard croaking to the echo of their own voices.

It has been proven that the eyeless fish prey upon each other. In shape they somewhat resemble the common catfish, and rarely exceed eight inches in length. One of these fish was caught by the guide in our presence, placed in a bottle of water, and taken out of the Cave alive, and might have been brought home with us, without impairing its vitality, if supplied daily with fresh water. They are captured by means of a small scoop net, which is gently carried beneath them.

Professor B. Silliman, Jr., who visited the Cave in the autumn of 1850, published the following observations on the blind fish and the blind crawfish, in "Silliman's Journal" for May, 1851:

"Of the fish there are two species, one of which has been described by Dr. Wyman in the 'American Journal of Science,' and which is entirely eyeless; some ten or twelve specimens of the species were obtained. The second species of the fish is not colorless like the first, and it has external eyes, which, however, are found to be quite blind. The crawfish, or small crustacea, inhabiting the rivers with the fish, are also eyeless, and uncolored; but the larger eyed and colored crawfish, which are abundant within the Cave, are also common, at some seasons, in the subterranean rivers, and so also, it is said, the fish of Green River are to be found at times of flood in the rivers of the Cave. Among the collections are the larger eyed crawfish, which were caught by us in the Cave."

For the benefit of those who may feel interested in the scientific characters and peculiarities of the Cave fish, we will quote the observations of two authorities, whose names are generally recognized and respected in scientific circles. In " Silliman's Journal" for January, 1851, p. 127, Professor Louis Agassiz, perhaps the most eminent living naturalist, especially in the department of ichthyology, in reply to a letter of inquiry from the senior editor of the "Journal," remarks, "The blind fish of the Mammoth Cave was for the first time described in 1812, in the Zoology of New York, by Dr. Dekay, Part 3d, page 187, under the name of 'Amblyopsis spelaeus; and referred, with doubt, to the family of Siluridae, on account of a remote resemblance to my genus Cetopsis. Dr. J. Wyman has published a more minute description of it, with very interesting anatomical details, in vol. xlv. of the 'American Journal of Science and Arts,' 1843, page 94.

"In 1844 Dr. Tellkampf published a more extended description, with figures, in 'Muller's Archiv ' for 1844, and mentioned several other animals found also in the Cave, among which the most interesting is a Crustacean, which he calls 'Astacus pellucidus,' already mentioned, but not described, by Mr. Thompson, President of the Natural History Society of Belfast. Both Thompson and Tellkampf speak of eyes in these species; but they are mistaken. I have examined several specimens, and satisfied myself that the peduncle or the eye only exists; but there are no visible facets at its extremity, as in other crawfish. (Speaking of the eyes of animals, it is remarked in the valuable school book of Professors Agassiz and Gould, entitled " Principles of Zoology," Boston, 1859. "Others, which live in darkness, have not even rudimentary eyes, as, for example, that curious fish *(Amblyopsis spelaeus)* which lives in the Mammoth Cave, and which appears to want even the orbital cavity. The crawfishes *(Astacus pellucidus)* of this same Cave are also blind, having merely the pedicle for the eyes, without even traces of facettes.")

"Mr. Thompson mentions, further, crickets, allied to Phalangopsis longipes,' of which Tellkampf says that it occurs throughout the Cave. Of spiders, Dr. Tellkampf found two eyeless, small, white species, which he calls 'Phalangodes armata' and 'Anthrobia monmouthia' flies, of the genus 'Anthomyia' a minute shrimp, called by him 'Triura cavernicola,' and two blind beetles 'Anophthalmus Tellkampfii' of Erichson, and 'Adelops hirtus;' of most of which Dr. Tellkampf has published a full

66

description and figures in a subsequent paper, inserted in Erichson's Archiv, 1844, p. 318.

"The infusoria, observed in the Cave resemble 'Monas Kolpoda," Monas socialis,' and 'Bodo intestinalis' a new Chilomonas, which he calls 'Ch. emarginata,' and a species allied to 'Kolpoda cucullus.'

"As already mentioned, Dekay has referred the blind fish, with doubt, to the family of Siluridae Dr. Tellkampf, however, establishes for it a distinct family. Dr. Storer, in his Synopsis of the Fishes of North America, published in 1846, in the Memoirs of the American Academy of Arts and Sciences, is also of opinion that it should constitute a distinct family, to which he gives the new name of Hypsaeidae, page 435. From the circumstance of its being viviparous, from the character of its scales, and from the form and structure of its head, I am inclined to consider this fish as an aberrant type of my family of Cyprinodonts.

"You ask me to give my opinion respecting the primitive state of the eyeless animals of the Mammoth Cave. This is one of the most important questions to settle in Natural History, and I have, several years ago, proposed a plan for its investigation, which, if well conducted, would lead to as important results as any series of investigations which can be conceived; for it might settle, once and forever, the question, in what condition and where the animals now living on the earth were first called into existence. But the investigation would involve such long and laborious researches, that I doubt whether it will ever be undertaken. It has occurred to me that the final step would be a thorough anatomical study of the species found in the Cave, with extensive comparison of allied species found elsewhere, next, an investigation of the embryology of all of them, and, when fully prepared by such researches, an attempt to raise embryos, of the species found in the Cave, under various circumstances, different from those in which they are naturally found at present.

"If physical circumstances ever modified organized beings, it should be easily ascertained here. For my own part, however, I think that the blind animals of the Cave would only show organs of vision during their embryonic state, in conformity with the normal development of the respective types to which they belong, and that even when placed under a moderate influence of light, incapable of injuring them, but sufficient to favor the growth of their eyes in the allied species provided with them, the young of those species peculiar to the Cave would gradually grow blind, while the others would acquire perfect eyes; for I am convinced, from all I know of the geographical distribution of animals, that they were created under the circumstances in which they now live, within the limits over which they range, and with the structural peculiarities which characterize them at the present day. But this is a mere inference, and whoever would settle the question by direct experiment might be sure to earn the everlasting gratitude of men of science. And here is a great aim for the young American Naturalist who would not shrink from the idea of devoting his life to the solution of one great question."

It will be seen from the foregoing that Professor Agassiz maintains the opinion that the fish and the crawfish of the Cave, with their structural peculiarities, "were created under the circumstances in which they now live," but, as he very frankly adds, "this is a mere inference."

We have already stated that the inference drawn from our own observations and reflections is that these animals were originally supplied with the organs of vision; and since the above was penned we have noticed that Professor Joseph Jones, of Nashville, Tenn., has incidentally corroborated our view while treating the subject of Albinism. Referring to the effects of continued darkness upon various animals. Dr. Jones remarks:

"After extended investigation and examination of thousands of living specimens, I have never observed an albino among cold blooded animals.

"When this class of animals have been confined in dark caves, and excluded from the action of light, they present the appearance of the albino; and it is probable that, if the negro children, which are almost white at the time of their birth, were reared in total darkness, they would in like manner be white.

"I have seen living sirens from the caves of Africa, without a particle of coloring matter in their skins, and so transparent that the form and pulsations of the heart and the circulation of the blood could be discerned through the walls of the abdomen and chest; and Dr. Blackie has informed me that he has seen similar colorless salamanders in the dark caves of Northern Georgia.

"I have in my possession specimens of the blind fish *(Amblyopsis spelaeus),* the blind crayfish *(Astacus pellucidus),* and of the crickets with eyes, of the dark caverns of the caves of Kentucky, which are entirely wanting in coloring, resembling albinos. The absence of the ball from the socket of the eye in the blind fish, and the absence of the eye from the peduncles of the blind crayfish, may be most philosophically attributed to the absence of that agent upon which the production of color depends. And it is now well established that we may arrest and alter the development of the tadpole, and other animals, by raising the amount of physical forces, heat and light." *Observations and Researches on Albinism in the Negro Race.* By Joseph Jones, M.D., Professor of Physiology and Pathology in the Medical Department of the University of Nashville, Tennessee. Published in the Transactions of the American Medical Association, vol. xx., 1869, pp. 703, 704.

Having now given Prof. Agassiz's general views upon the natural history of the eyeless animals of the Cave, we will proceed to the anatomical construction of the organs of sight and hearing of the blind fishes, in the words of Professor Jeffries Wyman, which we take from Silliman's "American Journal of Science and Arts" for March, 1851, page 228. He says:

"The general structure of the blind fishes was described in a former number of this journal (July, 1843), but a more complete description was

70

given in the 'New York Journal of Medicine' by Tellkampf, who, in company with J. Miller, of Berlin, for the first time detected the existence of rudimentary eyes. (New York Journal of Medicine, vol. v. p. 84. 1845. Dr. Dekay had previously mentioned the existence of eyes, but was evidently misled by some other appearance, since he states that eyes exist of the usual size, but are covered by the skin. He had not dissected them.—Fauna of New York."—(Note by Prof. Wyman.)) They are described as one twelfth of a line in diameter, round, black, destitute of a cornea, having an external layer of pigment, beneath which is a colorless membrane. No nerve was detected in connection with the eye, and the contents of the globe were not determined with certainty. Professor Owen has described the organ as a simple eye speck, as in the leech, consisting of a minute tegumentary follicle, coated by dark pigment which receives the end of a special cerebral nerve. Dr. John C. Dalton, Jr., has also detected the eyes, and describes them as minute globular sacs containing blackish pigment, deeply imbedded in the adipose tissue of the orbit, and measuring a little less than one-seventy-eighth of an inch.

"Through the kindness of Mr. Charles Dean, of Cambridge, and of Professor Agassiz, I have had placed at my disposal some specimens of Amblyopsis, well preserved in alcohol, and have been able to make, in some respects, a more complete description than has yet been given. I have also had an opportunity of inspecting, superficially, fourteen specimens, varying from one inch and a half to four inches and a half in length, but in three or four only could the eyes be detected through the skin. In the three specimens recently dissected, the eyes were exposed only after the removal of the skin and the careful separation from them of the loose areolar tissue which fills the orbit. In a fish four inches in length, the eyes measured one sixteenth of an inch in their long diameter, were of an oval form, and black. A filament of nerve [the professor here refers to a diagram of the eye, which accompanies his paper] was distinctly traced from the globe to the cranial walls; but the condition of the contents of the cranium, from the effects of the alcohol, was such as to render it impracticable to ascertain the mode of connection of the optic nerve with the optic lobes. A few muscular fibres were traced to the immediate neighborhood of the eye, and even in contact with it, but were not ascertained to have that regular arrangement which is seen in the more completely formed eyes of other fishes.

"Examined under the microscope with a power of about twenty diameters, the following parts were satisfactorily made out: first, externally an exceedingly thin membrane, which invested the whole surface of the eye, and appeared to be continuous with a thin membrane covering the optic nerve, and which was therefore regarded as a sclerotic; second, a layer of pigment cells, for the most part of a hexagonal form, and which were most abundant about the anterior part of the eye; third, beneath the pigment a single layer of colorless cells, larger than a pigment cell, and each cell having a distinct nucleus; fourth, just in front of the globe a lenticular shaped transparent body, which consisted of an external membrane containing numerous cells with nuclei; this lens shaped body seemed to be retained in its place by a prolongation forward of the external membrane of the globe; fifth, the globe was invested by loose areolar tissue, which adhered to it very generally, and in some instances contained yellow fatty matter, in one specimen it formed a round spot, visible through the skin, on each side of the head, which had all the appearance of a small eye, its true nature was determined by the microscope only. It is not improbable that the appearance just referred to may have misled Dr. Dekay, where he states that the eye exists of the usual size, but covered by the skin.

"If the superficial membrane above noticed is denominated correctly the sclerotic, then the pigment layer may be regarded as the representative of the choroid. The form as well as the position of the transparent nucleated cells within the choroid correspond, for the most part, with the retina. All of the parts just enumerated are such as are ordinarily developed from and in connection with the encephalon, and are not in any way dependent upon the skin. But if the lenticular shaped body is the true representative of the crystalline lens, it becomes difficult to account for its presence in Amblyopsis according to the generally recognized mode of its development (since it is usually formed from an involution of the skin), unless we suppose that after the folding of the skin had taken place in the embryonic condition the lens retreated from the surface and all connection with the integument ceased.

"According to Quatrefages, however, the eye of Amblyopsis is contained wholly in the cavity of the dura mater, and yet it has all the appearance of being provided with a lens. If his description be correct, then the mode of development as well as the morphology of the eye in this remarkable fish is different from that of most other vertebrates, since the lens never could have been formed from an involution of the skin, nor could the eye, with its lens, as Professor Owen asserts, be a modified cutaneous follicle. That there should be different modes of development of parts of the eye in different animals is by no means improbable, since we find this actually to be the case in another organ of sense, the nose. In some fishes the nostrils result from a depression or involution of the skin simply, and do not at any period communicate with the mouth; while in all the higher vertebrates they are formed by subdivision of the primitive oral cavity. It is possible, therefore, that in Amblyopsis the lens may have been developed where we find it, and that it was never connected with the integument. Whatever views be taken with regard to its development, the anatomical characters which have been enumerated show that, though quite imperfect as we see it in the adult, it is constructed after the type of the eyes in other vertebrates. It certainly is not adapted to the formation of images, since the common integument and the areolar tissue which are interposed between it and the surface would prevent the transmission of light to it except in a diffused condition. No pupil, nor anything analogous to an iris, was detected, unless we regard as representing the latter the increased number of pigment cells at the anterior part of the globe."

In continuation of the same observations, the professor next treats of the Ear; and his remarks on this point are conclusive in contradiction of the sensational newspaper article previously mentioned. He says:

"It is said that the blind fishes are acutely sensitive to sounds, as well as to undulations produced by other causes in the water. In the only instance in which I have dissected the organ of hearing (which I believe has not been before noticed), all its parts were largely developed, as will be seen by reference to figures 2 and 3. [We regret that we are unable to introduce the figures here.] As regards the general structure, the parts do not differ materially from those of other fishes, except for their proportional dimensions. The semicircular canals are of great length, and the two which unite to enter the vestibule by a common duct, it will be seen, project upwards and inwards under the vault of the cranium, so as to approach quite near to the corresponding parts of the opposite side. The otolite contained in the utricle was not remarkable, but that of the vestibule is quite large when compared with that of a Leuciscus of about the same dimensions as the blind fish here described."

After these dry scientific details of some of the inhabitants of the Cave; which may prove interesting to those readers whose studies lead them in such channels, and for which we ask the forbearance of the general reader, we will proceed with the narrative of our journey.

# CHAPTER IX. - SILLIMAN'S AVENUE.

Upon landing on the farther shores of Echo River, we immediately enter Silliman's Avenue, so named for Professor B. Silliman, Jr., who visited the Cave in 1850.

This avenue is a mile and a half long, and extends from Echo River to the Pass of El Ghor. Its height varies from twenty to forty feet, and it is from twenty to two hundred feet in width. The walls and ceiling of this avenue are rugged and water worn. It is undoubtedly of recent formation, as compared with other parts of the Cave.

The objects of interest in Silliman's Avenue come in the following order:

1. CASCADE HALL is two hundred feet in diameter, and twenty feet high. It receives its name from a small cascade that falls into it from the ceiling. Of this hall Bayard Taylor says, "A few minutes of rough travel brought us to a large circular hall, with a vaulted ceiling, from the center of which poured a cascade of crystal water, striking upon the slant side of a large reclining boulder, and finally disappearing through a funnel shaped pit in the floor. It sparkled like a shower of pearls in the light of our lamps as we clustered around the brink of the pit to drink from the stores gathered in those natural bowls which seem to have been hollowed out for the uses of the invisible gnomes."

The avenue which leads to Roaring River takes its origin in Cascade Hall.

2.   THE DRIPPING SPRING is a pool of water that is supplied from the ceiling. Stalactites and stalagmites are found at this point.

3.   THE INFERNAL REGION receives its name from the fact that the floor is composed of wet clay and is exceedingly irregular. It is almost impossible to pass over it without receiving a fall. Several of our party would be willing to testify to this assertion.

4.   THE SEA SERPENT is a tortuous crevice in the rock overhead, that has been cut by running water, the layer of rock that formed the floor of it having been detached.

5.   THE VALLEY WAYSIDE CUT is a small avenue leading off from Silliman's Avenue and returning into it a short distance farther on. It presents several beautiful points, and is worth exploring.

6.   THE HILL OF FATIGUE is appropriately named, being hard to climb, but is not otherwise worthy of note.

7.   THE GREAT WESTERN, so called from the supposition that in appearance it represents the immense ship known as the Great Eastern, is an enormous rock, many times larger than any vessel, the end of which closely resembles the stern of a ship. To make use of nautical language the rudder is turned to the starboard side.

8.   THE RABBIT is a large stone, which is supposed to resemble the animal whose name it bears.

9.   OLE BULL'S CONCERT ROOM is situated to the left of the avenue. It is thirty feet wide, forty long, and twenty high. When the great

violinist made his first tour through the United States, he visited the Cave and performed in the room which has received his name.

At the end of Silliman's Avenue begins RHODA'S ARCADE, which arises half a mile from the Pass of El Ghor; it is five hundred yards in length, and from five to ten feet in height.

The walls and ceiling are incrusted with crystals of gypsum and carbonate of lime, of great brilliancy and indescribable beauty. The floor is covered with white crystals of limestone, and is unobstructed by fallen rock. In point of beauty there is no avenue superior to this.

LUCY'S DOME is reached by passing through Rhoda's Arcade. It is about sixty feet in its greatest diameter, and over three hundred in height, being the highest dome in the Cave. The sides appear to be composed of immense curtains, extending from the ceiling to the floor.

We next reach THE PASS OF EL GHOR, which resembles Silliman's Avenue, but the cliffs composing its walls present a more wild and rugged appearance. It is about two miles in length.

Of this Pass, Bayard Taylor remarks that he supposes it was named by some traveler who had been in Arabia Petraea, and adds that the name is a pleonasm, as el ghor signifies a narrow, difficult pass between rocks.

Mr. Taylor regarded the Pass of El Ghor as by far the most picturesque avenue in the Cave. He continues: "It is a narrow, lofty passage meandering through the heart of a mass of horizontal strata of limestone, the broken edges of which assume the most remarkable forms. Now there are rows of broad, flat shelves overhanging your head; now you enter a little vestibule with friezes and mouldings of almost Doric symmetry and simplicity; and now you wind away into a Cretan Labyrinth, most uncouth and fantastic, whereof the Minotaur would be a proper inhabitant. It is a continual succession of surprises, and, to the appreciative visitor, of raptures."

We will specify the objects of interest in this avenue as they present themselves:

1. THE HANGING ROCKS look as though they were on the point of falling and closing the avenue over which they are suspended; but, as before stated, no rocks from the walls or ceiling have been known to fall in any part of the Cave since its discovery.

2. THE FLY CHAMBER receives its name from the fact that crystals of black gypsum, of the size of a common house fly, project in great numbers from the ceiling

3. TABLE ROCK is twenty feet long, and projects from the left side of the avenue about ten feet. It is about two feet in thickness.

4. THE CROWN is six feet in diameter, and is situated on the right side of the avenue, about ten feet from the floor. It closely resembles the object after which it is named.

5. BOONE'S AVENUE leads off to the left. It has been explored only about one mile, and nothing further is known as to its extent or dimensions.

6. CORINNA'S DOME rests directly over the center of the avenue. It is forty feet high and nine wide. It was formed by the solvent action of

water, which entered it through a fissure at the top, when the Pass of El Ghor was filled with water. Had it been formed after the water had left the avenue, there would have been a pit beneath it, as at Shelby's Dome and the Bottomless Pit.

7. THE BLACK HOLE OF CALCUTTA is situated on the left side of the avenue, and is about fifteen feet deep.

8. STELLA'S DOME is two hundred and fifty feet in height, and, in general appearance, resembles Lucy's Dome. It is reached by passing through a small avenue which enters the left wall of the Pass of El Ghor.

9. THE CHIMES consist of depending rocks, as in the Bacon Chamber, which, when struck, emit a musical sound. These are objects of interest to the visitor.

10. WELLINGTON'S GALLERY is not attractive.

11. HEBE'S SPRING is about four feet in diameter, and one foot and a half in depth. Its water is charged with sulphuretted hydrogen. Twenty years ago, as we are informed by Dr. Wright, there was no sulphur in this Spring, and at the present time, when it has been undisturbed for several hours, pure water is found upon the surface, and sulphur water at the bottom, indicating the fact that it is supplied with sulphur water at the bottom and pure water near the surface which come from entirely different sources. Eyeless crawfish have been found in Hebe's Spring.

At the distance of half a mile beyond Hebe's Spring, the Pass of El Ghor communicates with a body of water, the extent of which is unknown, called the Mystic River.

# CHAPTER X. - MARTHA'S VINEYARD.

We leave the Pass of El Ghor at the foot of Martha's Vineyard.

The avenue which contains Martha's Vineyard is elevated twenty feet above the Pass of El Ghor, and is reached, with considerable difficulty, by ascending a steep ladder near Hebe's Spring.

Of this curiosity, Bayard Taylor expressed his impressions in the following language: "We were now, according to the guide's promises, on the threshold of wonders. Before proceeding farther, we stopped at Hebe's Spring, which fills a natural basin in the bottom of a niche made on purpose to contain it. We then climbed a perpendicular ladder, passing through a hole in the ceiling barely large enough to admit our bodies, and found ourselves at the entrance of a narrow, lofty passage leading upward. When all had made the ascent, the guides exultingly lifted their lamps and directed our eyes to the rocks overhanging the aperture. There was the first wonder, truly! Clusters of grapes, gleaming with blue and violet tints through the water which trickled over them, hung from the cliffs, while a stout vine, springing from the base and climbing nearly to the top, seemed to support them. Hundreds on hundreds of bunches, clustering so thickly as to conceal the leaves, hang, forever ripe and forever unplucked, in that marvelous vintage of the subterranean world. For whose hand shall squeeze the black, infernal wine from the grapes that grow beyond Lethe?"

*Grape Clusters*

Dr. Wright tells us, in more sober language, that the walls and ceiling of Martha's Vineyard are studded with stalactite nodules of carbonate of lime, colored with the black oxide of iron, and in size and appearance resembling grapes. A stalactite three inches in diameter, and extending from the floor to the ceiling, is termed the Grape Vine.

A large stalagmite projects from the right wall, a few inches from the floor, and is termed the Battering Ram.

Elindo Avenue takes its origin directly over the Pass of El Ghor. It presents no points of special interest, except that the avenue leading to the Holy Sepulchre, which is situated directly over Martha's Vineyard, and which contains a fine collection of stalactites, arises in it. In speaking of this part of the Cave, Mr. Martin remarks:

"About one hundred feet from this spot, taking the right, over a rough and rather a difficult way, the tourist at last reaches what is called the height or hill. On this stands the Holy Sepulchre. This natural chapel is about twelve feet square; it has a low ceiling, and is decorated in the most magnificent style imaginable, having well arranged draperies of stalactite of every possible shape. You go to the room of the Holy Sepulchre, adjoining. Unlike the place you left, it is without ornament or decoration of any kind whatever; it presents nothing but dark and bare walls, and has been likened, by many who have been there, to a charnel house. In the center of this room, which stands but a few feet below the chapel, the visitor will be shown what seems to be a grave hewn out of the solid rock. So great is the resemblance as to have suggested to a Roman Catholic priest the exclamation that has since passed as its name. The reverend gentleman referred to no sooner cast his eyes upon this opening in the rock, than he cried out, on bended knees and with uplifted hands, "The Holy Sepulchre! The Holy Sepulchre!"

Continuing our advance, we next arrived at a point of great interest, a locality that had been anxiously inquired for, for more than an hour previously to reaching it, Washington Hall the place of dining.

Dr. Wright says that this Hall is generally reached between twelve and one o'clock; but our party, being composed of slow travelers, and of persons anxious to see at leisure everything of interest, did not arrive at the Hall until half past two. The following are Bayard Taylor's impressions of the dining apartment:

"Mounting for a short distance, this new avenue suddenly turned to the left, widened, and became level. The ceiling is low, but beautifully vaulted, and "Washington's Hall, which we soon reached, is circular, and upwards of one hundred feet in diameter. This is the usual dinner room of parties who go beyond the rivers. Nearly five hours had now elapsed since we entered the Cave, and five hours spent in that bracing, stimulating atmosphere might well justify the longing glances which we cast upon the baskets carried by the guides. Mr. Miller [the then proprietor of the hotel] had foreseen our appetites, and there were stores of venison, biscuit, ham, and pastry, more than sufficient for all. We made our mid day, or rather midnight, meal, sitting, like the nymph who wrought Excalibur, 'Upon the hidden bases of the hills,' buried far below the green Kentucky forests, far below the forgotten sunshine. For in the Cave you forget that there is an outer world somewhere above you. The hours have no meaning. Time ceases to be; no thought of labor, no sense of responsibility, no twinge of conscience, intrudes to suggest the existence you have left. You walk in some limbo beyond the confines of actual life yet no nearer the world of spirits. For my part, I could not shake off the impression that I was wandering on the outside of Uranus, or Neptune, or some planet still more deeply buried in the frontier darkness of our solar system."

We indorse all that we have quoted from Mr. Taylor.

"There may be," remarked our corpulent friend B., "a great deal of romance in this way of eating, with your plate on your lap and seated on a rock or lump of nitre earth; but, for my part, I would rather dispense with the poetry of the thing, and eat a good dinner, whether above or below ground, from off a bona fide table, and seated on a good substantial chair. The proprietor ought to have, at all the dining places, tables, chairs, and the necessary table furniture, that visitors might partake of their collations with some degree of comfort."

We regard this as a very proper suggestion. The proprietor of the hotel might, with very trifling cost, keep permanently, at the two or three principal places of dining, substantial tables and chairs for the accommodation of visitors.

Cans of oil are kept in this room (Washington Hall), from which the lamps are replenished. Although the lamps are capable of holding oil sufficient to burn ten hours, the depots for it are so arranged that they can be filled every five hours; and, as a greater security against total darkness, the guide carries a bottle of oil in his satchel.

Marion's Avenue, which rises in Washington Hall, leads to Paradise, Zoe's Grotto, and Portia's Parterre. These avenues will be again referred to.

Upon leaving Washington Hall, and before reaching Cleveland's Cabinet, we pass through the Snowball Room. The ceiling is studded with white nodules of gypsum, which vary in diameter from two to four inches. The atmosphere of the room is too damp for the gypsum to assume the forms of flowers and filaments, as it does in Cleveland's Cabinet. The resemblance of these nodules to snowballs is complete.

# CHAPTER XI. - CLEVELAND'S CABINET AND THE ROCKY MOUNTAIN.

We now enter the last avenue on the "Long Route," which, in point of attractiveness and extraordinary beauty, is the crowning glory of the Cave. We refer to Cleveland's Cabinet. This avenue is about two miles in length, extending to the Rocky Mountain. The interest connected with this avenue is so great that we hope we will be pardoned for here inserting extended extracts from the observations of others regarding the impressions produced upon them while witnessing its curiosities.

Mr. Martin says, "This avenue is truly magnificent; it may be designated one of the most magnificent objects in the world. It is replete with formations that are to be seen in no other place, which even the dullest cannot behold without experiencing sensations quite new to them, but which in the cultivated and intellectual awaken feelings of rapture.

"Professor Locke has designated some of these formations as onlophilites, or curled leaf stones. In lecturing on them he says, 'They are unlike anything yet discovered, equally beautiful for the cabinet of the amateur and interesting to the geological philosopher.'

"Another gentleman (a clergyman) also speaks of these formations. His remarks are to the following effect: So exquisite and beautiful is Cleveland's Avenue, that it is out of the power of painter or poet to conceive anything like it. Such loveliness cannot, indeed, be described. Were the sovereigns of wealthy states to spend their all on the most skillful lapidaries they could find, with the view of rivaling the splendor of this truly regal abode, the attempt would be entirely vain. What, then, is left for the narrator? People must see it, and then they will be convinced that all attempts at adequate description are useless. The Cabinet was discovered by Mr. Patten, of Louisville, and Mr. Craig, of Philadelphia, accompanied by the guide Stephen. It extends in a direct line about two miles. It is a perfect arch of fifty feet span, and of an average height of ten feet in the center, just high enough to be viewed at ease in all its parts. It is incrusted from end to end with the most beautiful formations in every variety of form. The base of the whole is sulphate of lime, in one part of dazzling whiteness and perfectly smooth, and in other places crystallized so as to glitter like diamonds in the light. Growing from this, in endless diversified forms, is a substance resembling selenite, translucent and imperfectly laminated. Some of the crystals bear a striking resemblance to celery, and all are about the same length, while others, a foot or more in length, have the color and appearance of vanilla cream candy; others are set in sulphate of lime, in the form of a rose; and others still roll out from the base in forms resembling the ornaments on the capital of a Corinthian column. Some of the incrustations are massive and splendid; others are as delicate as the lily, or as fancy work of shell or wood. Let any person think of traversing an arched way like this for two miles, and all the wonders of the tales of youth, not forgetting those gorgeous fictions, The Arabian Nights, seem tame and uninteresting when brought into comparison with the living, growing reality. The term 'growing' is not a misnomer; the process is going on before your eyes. Successive coats of these incrustations have been perfected, and then crowded off by others, so that hundreds of tons of these gems lie at your feet and are crushed as you pass, while the work of restoring the ornaments for Nature's boudoir is proceeding around you. Here and there, through the whole extent, you will find openings through the side, into which you may thrust the person and often stand erect in little grottos, perfectly incrusted with a delicate white substance, reflecting the light from a thousand glittering points. Many visitors are so enraptured with the place that they cannot repress exclamations of surprise and worship."

This beautiful avenue (Cleveland's) is of sufficient importance to justify us in giving the views of various authors with regard to it. We will next add an extract from the sketch of Bayard Taylor:

"It is completely incrusted from end to end with crystallizations of gypsum, white as snow. This is the crowning marvel of the Cave, the pride and the boast of the guides. Their satisfaction is no less than yours, as they lead you through the diamond grottos, the gardens of sparry efflorescence, and the gleaming vaults of this magical avenue. We first entered the Snowball Room, where the gnome children, in their sports, have peppered the gray walls and ceiling with thousands of snow white projecting disks, so perfect in their fragile beauty that they seem ready to melt away under the blaze of your lamp. Then commences Cleveland's Cabinet, a gallery of crystals, the richness and variety of which bewilder you. It is a subterranean conservatory, filled with the flowers of all zones; for there are few blossoms expanding on the upper earth but are mimicked in these gardens of darkness. I cannot lead you from niche to niche, and from room to room, examining in detail the enchanted growths; they are all so rich and so wonderful that the memory does not attempt to retain them. Sometimes the hard limestone rock is changed into a parterre of white roses; sometimes it is starred with opening daisies; the sunflowers spread their flat disks and rayed leaves; the feathery chalices of the cactus hang from the clefts; the night blooming cereus opens securely her snowy cup, for the morning never comes to close it; the tulip is here a virgin, and knows not that her sisters above are clothed in the scarlet of shame.

"In many places the ceiling is covered with a mammillary crystallization, as if a myriad bubbles were rising beneath its glittering surface. Even on this jeweled soil, which sparkles all around you, grow the lilies and roses, singly overhead, but clustering together toward the base of the vault, where they give place to long, snowy, pendulous cactus flowers, which droop like a fringe around diamonded niches. Here you see the passion flower, with its curiously curved pistils; there an iris, with its lanceolate leaves; and again, bunches of celery, with stalks white and tender enough for a fairy's dinner. There are occasional patches of gypsum, tinged with a deep amber color by the presence of iron. Through the whole length of the avenue there is no cessation of the wondrous work. The pale rock-blooms burst forth everywhere, crowding on each other until the brittle sprays cannot bear their weight, and they fall to the floor. The slow, silent efflorescence still goes on, as it has done for ages in that buried tropic.

"What most struck me in my underground travels," continues Mr. Taylor, "was the evidence of design which I found everywhere. Why should the forms of earth's outer crust, her flowers and fruits, the very heaven itself which spans her, be so wonderfully reproduced? What law shapes the blossoms and the foliage of that vast crystalline garden? There seemed to be something more than the accidental combinations of a blind chance in what I saw, some evidence of an informing and directing will. In these secret caverns, the agencies which produced their wonders have been at work for thousands of years, perhaps thousands of ages, fashioning the sparry splendors in the womb of darkness with as exquisite a grace, as true an instinct of beauty, as in the palm or the lily, which are moulded by the hands of the sun. What power is it that lies behind the mere chemistry of Nature, impregnating her atoms with such subtle laws of symmetry? What but the Divine will which first gave her being, and which is never weary of multiplying for man the lessons of infinite wisdom?"

Professor Locke, of the Ohio Medical College, under date of Cincinnati, October 26, 1841, published in Silliman's "American Journal of Science and Arts" for 1842, p. 206, a sketch entitled "Alabaster in the Mammoth Cave of Kentucky," from which we make the following extracts:

"After crossing within the Cave several streams in boats, an apartment has been reached (Cleveland's Cabinet), the roof of which is decorated with the most gorgeous ornaments of alabaster, so much like a work of art as to surpass credibility. They are white and semi-transparent, and are thrown out from the rock in the form of rosettes, leaves, and curled enrichments of the composite capital of architecture. * * * I was at first at a loss to account for such beautiful formations, and especially for the elegance of the curves exhibited. It is, however, evident that the substances have grown from the rocks by increments or additions to the base, the solid parts already formed being continually pushed forward. If the growth be a little more rapid on one side than on the other, a well proportioned curve will be the result; should the action on one side diminish or increase, then all the beauties of the conic and mixed curves would be produced. The masses are often evenly and longitudinally striated by a kind of columnar structure, exhibiting a fascicle of small prisms, and some of these prisms, ending sooner than others, give a broken termination of great beauty, similar to our form of the 'emblem of the order of the Star.' The rosettes formed by a mammillary disk, surrounded by a circle of leaves at every flexure, like the branches of a chandelier, running more than a foot in length and not thicker than the finger, are among the varied frost work of the alabaster grottos; common stalactites of carbonate of lime, although beautiful objects, lose by contrast with these ornaments all of their effect, and dwindle into mere clumsy, awkward icicles."

*Rosa's Bower*

Having given a general idea of the beauty and grandeur of Cleveland's magnificent avenue, we proceed to mention in detail the objects of interest as they present themselves:

1. MARY'S BOWER is fifteen feet in height and forty in length. Its walls and ceiling are covered with rosettes of gypsum. Immediately adjoining Mary's Bower, and by many regarded as surpassing it in beauty, is to be seen ROSA'S BOWER, a very good representation of which has been produced by our artist.

2. THE CROSS consists of two crevices in the ceiling, intersecting each other at right angles, and lined with flowers of plaster of Paris. It is about eight feet in length.

3. THE MAMMARY CEILING is formed of nipple shaped projections of gypsum.

4. THE LAST ROSE OF SUMMER is about eight inches in diameter, and is of snowy whiteness. It rests against the ceiling, in the center of the avenue. This is indeed a marvel of beauty. It is a perfect representation of a very large rose. It hangs alone, a short distance beyond the thousands of clusters, and it is really the last to be found in the avenue.

5. THE DINING TABLE is fifteen feet wide and thirty long. It is a flat rock that has been detached from the ceiling.

6. BACCHUS'S GLORY is an alcove, three feet in height and five feet in length, the whole interior of which is lined with nodules of gypsum which in size and form resemble grapes. It is situated to the left of the Dining Table.

7.   ST. CECILIA'S GROTTO is remarkable for the size of the stucco flowers found in it.

8.   DIAMOND GROTTO is lined with crystals of selenite, which, when a light is waved to and fro in front of them, sparkle like the gem after which the grotto is named.

It is to be hoped that the beautiful but delicate formations in Cleveland's Cabinet will ever be carefully guarded against the destructive hand of man for by human agency all those wonderful rock-blooms which have occupied thousands of years in their production might be destroyed in a few hours. The Cabinet is named in honor of Professor Cleveland, the distinguished mineralogist.

At the termination of Cleveland's Cabinet we arrive at the base of the Rocky Mountain. This mountain is one hundred feet high, and is entirely formed of rocks that have fallen from above. On the top of the Rocky Mountain there is a stalagmite two feet high and six inches in diameter, termed Cleopatra's Needle.

On the farther side of the Rocky Mountain is a gorge seventy feet deep and one hundred feet wide, termed Dismal Hollow. The Cave at the Mountain divides into three branches. That to the right leads to Sandstone Dome, which is interesting from the fact that the stone of which it is composed indicates that the top of the Dome is near the surface of the earth. The branch to the left communicates with Crogan's Hall, named for Dr. Crogan, a former proprietor of the Cave. The central one is termed Franklin Avenue, and extends from Dismal Hollow to Serena's Arbor.

Franklin Avenue, as before stated, extends from Dismal Hollow to Serena's Arbor, a distance of a quarter of a mile. It varies in width from thirty to sixty feet; it has a wild and gloomy appearance.

Serena's Arbor is twenty feet in diameter and about forty in height. The walls and ceiling are covered with stalactite cornices, columns, grooves, ogees, etc., many of which are semitransparent and sonorous.

At the base of the Rocky Mountain the guide stopped, intimating that the terminus of the journey had been reached. Having read a thrilling account of a descent into the Maelstrom some years ago, we expressed a desire to see the awful pit, which was some distance beyond. All the gentlemen of the party, and one or two of the ladies also, expressed a willingness to climb the Rocky Mountain. The other ladies awaited our return. The ascent of the mountain was extremely difficult, and it is not to be wondered at that the guides do not insist upon visitors passing over it.

Beyond the mountain we enter Crogan's Hall, which constitutes the end of the Long Route, and which is about seventy feet wide and twenty high. The left wall is covered with stalactite formations, which are white and semi-transparent and of great hardness, and fragments of which are sometimes worked into ornaments.

# CHAPTER XII. - THE MAELSTROM. A PERILOUS ADVENTURE.

The Maelstrom is a pit one hundred and seventy five feet deep and twenty wide. There are avenues leading from the bottom, which may be seen when a light is lowered into it, but which have been, as yet, imperfectly explored.

In connection with the Maelstrom, we cannot refrain giving the graphic and thrilling account of the adventure, already alluded to, of William Courtland Prentice, son of George D. Prentice, the venerable editor of the "Louisville Journal," who was an officer in the Confederate Army, and was killed in a raid on the banks of the Ohio in 1862. In referring to his untimely death, the "Journal" said:

"He loved to seek the wildest and loneliest portions of Kentucky. Repeatedly he went far up among the bald and desolate crags of the cliffs of Dix River, a region haunted by the bear, the wildcat, and the catamount. The piercing scream of the panther even then was a sound of rapture to his ear. He was ever in search of natural curiosities, and he discovered and explored caves previously unknown, in all probability, to any man of our generation, and in one of them be found immense numbers of human bones that seemed to him to have belonged to a different order of beings from any now upon our continent. He subsequently became as familiar with the Mammoth Cave as the best of its guides. An adventure of his in that subterranean realm attracted much attention four years ago. An account of it was published in our columns, and, as we have often been requested to republish it, we will do so now:

"TERRIFIC ADVENTURE IN THE MAMMOTH CAVE. At the supposed end of what has always been considered the longest avenue in the Mammoth Cave, nine miles from its entrance, there is a pit, dark and deep and terrible, known as the Maelstrom. Tens of thousands have gazed into it with awe while Bengal lights were thrown down to make its fearful depths visible, but none had ever the daring to explore it. The celebrated guide Stephen, who was deemed insensible to fear, was offered six hundred dollars by the proprietors of the Cave if he would descend to the bottom of it; but he shrank from the peril. A few years ago, a learned and bold man resolved to do what no one before him had dared to do; and, making his arrangements with great care and precaution, he had himself lowered down by a strong rope a hundred feet, but at that point his courage failed him, and he called aloud to be drawn out. No human power could ever have induced him to repeat the appalling experiment.

"A couple of weeks ago, however, a young gentleman of Louisville (Wm. Courtland Prentice), whose nerves never trembled at mortal peril, being at the Mammoth Cave with Professor Wright, of our city, and others, determined, no matter what the dangers might be, to explore the depths of the Maelstrom. Mr. Proctor, the enterprising proprietor of the Cave, sent to Nashville, and procured a long rope of great strength expressly for the purpose. The rope and some necessary timbers were borne by the guides and others to the point of the exploration. The arrangements being soon completed, the rope, with a heavy fragment of rock affixed to it, was let down and swung to and fro to dislodge any loose pieces of rocks that would be likely to fall at the touch. Several were thus dislodged, and the long continued reverberations, rising up like distant thunder from below, proclaimed the depth of the horrid chasm. Then the young hero of the occasion, with several hats drawn over his head, to protect it as far as possible against masses falling from above, and with a light in his hand and the rope fastened around his body, took his place over the awful pit, and directed the half dozen men, who held the end of the rope, to let him down into the Cimmerian gloom.

"We have heard from his own lips an account of his descent. Occasionally masses of earth and rock went whizzing past, but none struck him. Thirty or forty feet from the top, a cataract from the side of the pit went rushing down the abyss, and, as he was in the midst of the spray, he felt some apprehension that his light would be extinguished; but his care prevented this. He was landed at the bottom of the pit, a hundred and ninety feet from the top. He found it almost perfectly circular, about eighteen feet in diameter, with a small opening at one point, leading to a fine chamber of no great extent. He found on the floor beautiful specimens of black selix, of immense size, vastly larger than were ever discovered in any other part of the Mammoth Cave, and also a multitude of exquisite formations as pure and white as virgin snow. Making himself heard, with great effort, by his friends, he at length asked them to pull him partly up, intending to stop on the way and explore a cave that he had observed opening about forty feet above the bottom of the pit.

Reaching the mouth of the cave, he swung himself with much exertion into it, and, holding the end of the rope in his hand, he incautiously let it go, and it swung out apparently beyond his reach. The situation was a fearful one, and his friends above could do nothing for him. Soon, however, he made a hook of the end of his lamp, and by extending himself as far over the verge as possible without falling, he succeeded in securing the rope. Fastening it to a rock, he followed the avenue one hundred and fifty or two hundred yards to a point where he found it blocked by an impassable avalanche of rock and earth. Returning to the mouth of this cave, be beheld an almost exactly similar mouth of another on the opposite side of the pit, but not being able to swing himself into it, he refastened the rope around his body, suspended himself again over the abyss, and shouted to his friends to raise him to the top. The pull was an exceedingly severe one, and the rope, being ill adjusted around his body, gave him the most excruciating pain. But soon his pain was forgotten in a new and dreadful peril.

"When he was ninety feet from the mouth of the pit, and one hundred from the bottom, swaying and swinging in mid air, he heard rapid and excited words of horror and alarm above, and soon learned that the rope by which he was upheld had taken fire from the friction of the timber over which it passed. Several moments of awful suspense to those above, and still more awful to him below, ensued. To them and to him a fatal and instant catastrophe seemed inevitable. But the fire was extinguished with a bottle of water belonging to himself, and then the party above, though almost exhausted by their labors, succeeded in drawing him to the top. He was as calm and self possessed as upon his entrance into the pit; but all of his companions, overcome by fatigue, sank down upon the ground, and his friend, Professor Wright, from overexertion and excitement, fainted, and remained for some time insensible.

"The young adventurer left his name carved in the depths of the Maelstrom, the name of the first and only person that ever gazed upon its mysteries."

The guides informed us, upon reference being made to this terrific adventure, which had previously come under our notice, that since the occasion of Prentice's descent, two other parties have been bold enough to incur the same hazards, one an Englishman and the other an American. They, however, did not meet with the appalling experience that is so graphically narrated in the foregoing sketch.

# CHAPTER XIII. - THE RATS, INSECTS, ETC. OF THE CAVE.

Having now reached the end of the Long Route, and presuming that the minds of our readers are sufficiently filled, for the present, with the beautiful, the sublime, and the terrible, we embrace the opportunity, before wending back our weary way to the land of verdure and sunshine, to draw attention for a few moments, by way of diversion, to some of the animals and insects of the Cave which have not already been described, several of which are found at this point.

A peculiar kind of rat is sometimes found in Crogan's Hall as well as in other parts of the Cave, which is a size larger than the Norway rat. The head and eyes resemble those of a rabbit, and the hair on the back is like that of a gray squirrel, but that of the legs and abdomen is white.

Cave crickets and lizards are also found here. The Cave crickets are about an inch in length.

The body is yellow, striped with black. They are provided with large eyes, but seem to direct their course, mainly, by their antennae, or feelers, which are enormously developed. They are sluggish in their movements, and, unlike other crickets, observe an eternal silence.

The Cave lizards vary in length from three to five inches. The eye is large and prominent. The body is yellow and dotted with black spots, and is semi-transparent. They are sluggish in their movements.

The abundance of animal life at this point (Crogan's Hall) would seem to indicate that there is a communication with the surface of the earth at no great distance.

Bats are found in all parts of the Cave, we are told by Dr. Wright, but most abundantly in Audubon's Avenue.

Professor Silliman, a portion of whose remarks have already been quoted ("Silliman's Journal" for May, 1851), says:

"There are several insects, the largest of which is a sort of cricket with enormously long antennae. Of this insect, numerous specimens will be found among the specimens sent to Professor Agassiz. There are several species of coleoptera, mostly burrowing in the nitre earth. There are some small water insects also, which I suppose are crustacean. Unfortunately, three vials, containing numerous specimens of these insects, were lost with my valise from the stagecoach, and I fear will not be recovered.

"The only mammal, except the bats, observed in the Cave, is a rat, which is very abundant, judging from the tracks which they make, but so shy and secluded in their habits that they are seldom seen. We caught two of them, and, fortunately, they were male and female.

"The chief points of difference from the common rat, in external characters, are in the color, which is bluish, the feet and belly and throat white, the coat, which is of soft fur, and the tail also thinly furred, while the common, or Norway rat, is gray or brown, and covered with rough hair. The Cave rat is possessed of dark, black eyes, of the size of a rabbit's eye, and entirely without iris; the feelers, also, are uncommonly long. We have satisfied ourselves that he is entirely blind when first caught, although his eyes are so large and lustrous."

We interrupt Professor Silliman here to suggest that the inability of the rat to see was perhaps owing to the unaccustomed, blinding light by which it was examined. It will be seen that the eye of the animal, gradually accustomed to light, finally becomes adapted to the new medium, and manifests the ability of exercising the sense of sight. This being the fact, it is to be inferred that the organs of vision were originally in a perfect condition, and afterward adapted to the state of darkness in which the animal existed; which may be conjectured to be a transitory state to a total obliteration of the visual organs, as has been accomplished in the fishes.

Professor Silliman continues:

"By keeping them [the rats], however, in captivity, and in diffuse light, they gradually appeared to attain some power of vision. They feed on apples and bread, and will not at present [soon after capturing them] touch animal food. There is no evidence that the Cave rats ever visit the upper air, and there was no one who could tell me whether they were or were not found there by the persons who first entered this place in 1802.

"Bats are numerous in the avenues within a mile or two of the mouth of the Cave, and Mr. Mantell thinks he has secured at least two species. Several specimens are preserved in alcohol. It was not yet quite late enough in the season when we were at the Cave, Oct. 16th- 22d, for all the bats to be in winter quarters, as the season was very open and warm. Still, in the galleries where they most abound, we found countless groups of them on the ceilings, chippering and scolding for a foothold among each other. On one little patch of not over four or five inches, we counted forty bats, and were satisfied that one hundred and twenty at least were able to stand on the surface of a foot square; for miles they are found in patches of various sizes, and a cursory glance satisfied us that it was quite safe to estimate them by millions. In these gloomy and silent regions, where there is neither change of temperature nor difference of light to warn them of the revolving seasons, how do they know when to seek again the outer air when the winter is over, and their long sleep is ended? Surely, He who made them has not left them without a law for the government of their lives."

It is supposed that the rats obtain their subsistence chiefly from the remnants of food left in various parts of the Cave by the visitors.

# CHAPTER XIV. - HOMEWARD BOUND.

We now turn our faces toward the outer world, the world which had, until this moment, been forgotten. We had been beguiled along from one scene of novelty and of grandeur and beauty to another; ever surprised and delighted with the endless variety, and mutely wondering what next would appear, until at length we found that we had reached the end of our journey, the "Ultima Thule" of the Cave, as Stephen was wont to say, without being conscious of bodily fatigue, and without remembering that we had already been eight hours away from that world where the sun shines, where the birds sing, and the fields display their verdure. But at the moment of turning back all these thoughts flood over us: a sense of physical weariness steals on us, and we are startled by the reflection that we are nine miles from the mouth of the cavern, and that there is no way of reaching it except by the same road over which we have already traveled, and by walking. Under such circumstances we feel a keen appreciation of the value of horses and railroads.

On our outward journey our party did not manifest the same degree of gayety or agility as during the inward passage. The contrast was striking: they filed along with heavy gait, and often in moody silence, it being frequently necessary for the leaders to halt, count heads, and drum up the stragglers. All appeared willing to take advantage of every opportunity to be seated without being fastidious as to the character of the seat.

On returning through Cleveland's Cabinet, all were anxious to secure specimens of the flowers, etc. The floor is strewn with fragments of these flowers, and visitors are privileged to take therefrom as many as they wish, without let or hindrance, but are not permitted, of course, to disturb those on the ceiling or walls. Many of the party selected the largest they could find. We, on the contrary, took pains to collect a variety of the smallest that would give a proper idea of their character, anticipating their oppressive weight on the long journey yet before us. Our conjectures proved correct. Before proceeding a mile several of the party began to throw down their specimens without a word of comment. The three or four pounds that were in our coat pockets became an intolerable burden upon the shoulders and neck. A young gentleman from Virginia, a Baptist minister, who accompanied our party, very accommodatingly proposed that we should place our specimens with his in his handkerchief, and carry them alternately. By this arrangement we succeeded between us in bringing out a very fair collection of some of the minor curiosities of the Cave.

It is unnecessary to refer to the incidents of our outward passage; they possess but little of interest. We made our exit from the Cave, footsore and weary, at half past nine o'clock in the evening, having been under ground for a period of twelve hours and a half. We climbed the rugged hill and reached the hotel, feeling as though it would have been impossible to have walked another step.

Though being exceedingly fatigued, for the reason that we had long been unaccustomed to much pedestrian exercise, yet we felt a thousand times repaid for our exhaustion in having visited a new world, and witnessed with our own eyes its multiform phases of wonder.

Awaiting our return, we found prepared a bountiful supper; and, happily, all were in a condition to do it full justice.

## CHAPTER XV. - THE SHORT ROUTE.

After the first day's underground journey it is decidedly better for the tourist to take at least twenty four hours repose, for the purpose of recuperating his physical forces, before undertaking further explorations. In our case our programme did not permit such a disposition of the time. Although we had slept soundly through the night, it was discovered on arising in the morning that our muscles were exceedingly stiff and sore. The greater number of our party, not having time at their disposal for further explorations, left for the North immediately after breakfast. In the mean time three or four new visitors had arrived. Our reduced party set out on the Short Route at half past nine o'clock. After exercising for a time, and witnessing new scenes, we soon forgot our corporeal fatigue.

We proceeded about one mile by the same route traveled the day previous, until we reached the Deserted Chamber. Here we left the Long Route, and, turning to the right, descended a pair of steps and entered the Labyrinth. This is a narrow, rugged causeway. The only object of interest to be found in it, says Dr. Wright, is the figure of the American Eagle on the left wall. The guide did not regard the resemblance of this figure to the bird, after which it is named, as very striking, but stated that he drew our attention to it for the reason that after leaving the Cave we would probably read in Dr. Wright's Manual that in this avenue was contained the figure of the American Eagle, and we would then censure the memory of the guide for not having pointed it out to us. We complimented the guide's views of ornithology, as also his conscientious discharge of duty.

Gorin's Dome, a curiosity of considerable magnitude, is reached by passing over a small bridge and ascending a ladder, ten feet in height, in the Labyrinth. It is viewed from a natural window, situated equidistant between the floor and the ceiling of the Dome. We are told by Dr. Wright that it is about two hundred feet in height and sixty feet across its widest part. The farther side presents a striking resemblance to an immense curtain, which extends from the ceiling to within forty feet of the floor. The window through which Gorin's Dome is viewed is circular in form, and not more, than two or three feet in diameter, allowing but one person at a time to enjoy the view of the interior. By imagining an immense well or deep circular excavation in the earth, without any opening at top or bottom, and supposing one's self to approach it about the center, or at a point midway between the floor and ceiling, and finding a small aperture through which a view of it can be obtained, would we not feel almost as much astonished at the novelty of the point of view as we would on beholding the curiosity itself?

Bayard Taylor speaks thus of this Dome:

"We now reached another pit, along the brink of which we walked, clambered up a ledge, and at last found a window-like opening, where Alfred (the guide) bade us pause. Leaning over the thin partition wall, the light of our united lamps disclosed a vast glimmering hall, the top of which vanished into darkness, and the bottom of which we could only conjecture by the loud, hollow splash of waterdrops that came up out of the terrible gloom. Directly in front of us hung a gigantic mass of rock, which, in its folds and masses, presented a wonderful resemblance to a curtain. It had a regular fringe of stalactites, and there was a short outer curtain overlapping it at the top. The length of this piece of limestone drapery could not have been less than one hundred feet. In a few moments, Alfred, who had left us, reappeared at another window on the right hand, where he first dropped some burning paper into the gulf, and then kindled a Bengal light. It needed this illumination to enable us to take in the grand dimensions of the Dome. We could see the oval arch of the roof a hundred feet above our heads; the floor studded with stalagmitic pedestals as far below; while directly in front the huge curtain that hung from the center of the Dome, the veil of some subterranean mystery, shone rosy white, and seemed to wave and swing, pendulous in the awful space. We were thoroughly thrilled and penetrated with the exceeding sublimity of the picture, and turned away reluctantly as the fires burned out, feeling that if the Cave had nothing else to show, its wonders had not been exaggerated."

Gorin's Dome was formed in the same manner as the Side Saddle Pit, which, it will be remembered, was by the solvent action of water charged with carbonic acid.

We are told that there are avenues which communicate with the top and bottom of the Dome. When Echo River rises, the floor of the Dome is covered with water, in which eyeless fish are sometimes found.

Gorin's Dome bears the name of its discoverer, and former proprietor of the Cave.

Pensacola Avenue is about a mile in length, from eight to sixty feet in height, and from thirty to one hundred in width. It is entered from Revelers' Hall.

The following objects are worthy of examination:

1.  THE SEA TURTLE is about thirty feet in diameter. The rock of which it is composed has fallen from the ceiling.

2.  THE WILD HALL in size and appearance resembles Bandit's Ball. Bunyan's Way, which communicates with Great Relief, enters Pensacola Avenue at this point.

3.  SNOWBALL ARCHED WAY receives its name from the fact that its ceiling is covered with nodules of gypsum, like those in the Snowball Room.

4.  THE GREAT CROSSING is the point at which four avenues take their origin.

5.  MAT'S ARCADE is fifty yards long, thirty feet wide, and sixty in height. Between the floor and ceiling there are four beautiful terraces, which extend the full length of the Arcade. There is a collection of beautiful stalactites, called the Pine Apple Bush, in Mat's Arcade.

6.  ANGELICO GROTTO, the ceiling and walls of which are incrusted with crystals of carbonate of lime, is a beautiful apartment. The artists have succeeded in obtaining a very handsome photograph of this grotto,

*Angelica's Grotto*

in the couch of which a young girl was reclining at the time of its execution. We and the following lines in the Rev. Horace Martin's book, addressed to this fairy grotto:

> Some Fairy of the olden time
> Her dwelling sure had here,
> And here she rested, with no grief
> To shade her spirits clear.
> How fit the place for one like her
> So fanciful and light,
> A creature jocund as the dawn,
> And as the morning bright.
> A spot like this, did he, the Bard
> Of Avon's flow'ry stream,
> Imagine, where, in 'frenzy fine,'
> He had his wanton 'Dream'.
> Titania here might move along,
> And Puck his frolics play,
> And Hermia, in her race of love,
> Outpace the hours of day.
> The fancy sees them passing now,
> How beautiful they seem!
> And now they're gone, we have them not,
> They vanish'd in the gloom."

Pensacola Avenue terminates about half a mile beyond Angelico Grotto, in a low archway.

Sparks' Avenue extends from River Hall to the Mammoth Dome, a distance of three fourths of a mile.

The objects of interest in this avenue occur in the following order:

1.    BANDITS' HALL is sixty feet long and forty wide, the floor of which is covered with large rocks that have been detached from the ceiling. It is truly a wild looking hall; its appearance naturally suggesting the name. Here bandits might retire, and hold their revels in perfect security. To the right of Bandits' Hall is an avenue of great extent, which has not been fully explored, called Brigg's Avenue.

2.    NEWMAN'S SPINE is about ten feet in length, and consists of a crevice in the center of the ceiling, which is the exact image of a cast of a gigantic backbone.

3.    SYLVAN AVENUE extends from Sparks' Avenue to Clarissa's Dome, and is about three hundred yards in length. This avenue contains a number of ferruginous logs, which vary from five to fifteen inches in diameter. Some of them appear to be chopped in half; others have lost a portion of bark, displaying a white surface of petrous wood; and others again look as though they were in a state of partial decay. Anywhere else these masses of stone might be mistaken for petrified wood.

Clarissa's Dome is entered at its base. It resembles Gorin's Dome, but is much smaller.

4.    BENNETT'S POINT is directly opposite Sylvan Avenue, where the avenue turns at an acute angle to the right. The floor of the avenue at this point is covered with yellow sand.

5.    BISHOP'S GORGE is a low and narrow part of the avenue, which is passed with difficulty.

Sparks' Avenue is named in honor of Mr. C. A. Sparks, of New York.

The Mammoth Dome is viewed from a terrace about forty feet from its base. It is two hundred and fifty feet in height, and in appearance closely resembles Gorin's Dome, but is more than five times as large. At the left extremity of the dome there are five large pillars cut out of the solid rock, called the Corinthian Columns.

The awful sublimity of this dome, when strongly illuminated, exceeds anything ever pictured to a mind frenzied by opium or hashish.

The Mammoth Dome is still enlarging.

The brief time that we were unfortunately restricted to when we visited the Cave, did not permit us to make a personal inspection of either Pensacola Avenue or Sparks' Avenue and Mammoth Dome, which we very much regret, and therefore are wholly indebted to Dr. Wright for the description which we have just given.

There are several avenues not often frequented by visitors, of which we need make no mention whatever, for the very good reason that we have no information to offer.

Roaring River is another portion of the Cave which we did not visit, but Dr. Wright informs us that the avenue which communicates with

Roaring River is entered at Cascade Hall, and is half a mile in length. He adds that Roaring River resembles Echo River in size and appearance, but has a louder echo. There is a cascade which falls into it, from which proceeds roaring sounds, and from which it has received its name.

Eyeless fish and eyeless crawfish are found in Roaring River, as well as sunfish and black crawfish, both of which are provided with eyes.

Marion's Avenue, Dr. Wright informs us, is about a mile and a half long, and arises in Washington Hall. It varies from twenty to sixty feet in width, and from eight to forty in height. The floor is covered with sand, and the walls are composed of white limestone, which resembles cumulus clouds. The far end of the avenue divides into two branches, that to the right leading to Paradise and Portia's Parterre, and that to the left to Zoe's Grotto.

The walls and ceiling of the avenue termed Paradise are covered with gypsum flowers. There is a dome in Paradise Avenue which is composed of sandstone. It is called Digby's Dome.

Portia's Parterre is entered from the left wall of Paradise Avenue. It is half a mile in length, and contains the same kind of flowers that are found in Cleveland's Cabinet. It was discovered about twelve years ago, and is commonly known as the New Discovery.

Upon leaving Gorin's Dome we returned to the Giant's Coffin, on our way to the Star Chamber.

The Star Chamber is situated in the Main Cave, which leads off to the left, as you enter, at the Giant's Coffin, as before mentioned. It is sixty feet in height, seventy in width, and about five hundred in length. The ceiling is composed of black gypsum, and is studded with innumerable white points, which, by a dim light, present a most striking resemblance to stars. These points, or stars, are produced, in part, by an efflorescence of Glauber's salts beneath the black gypsum, which causes it to scale off, and in part by throwing stones against it, by which it is detached from the white limestone. In the far extremity of the chamber a large mass has been separated, by which a white surface is exposed, termed the Comet.

We give below the observations and remarks of Bayard Taylor on approaching the Star Chamber, and his impressions upon witnessing it. He says:

"We passed several stone and frame houses, some of which were partly in ruin. The guide pointed them out as the residence of a number of consumptive patients, who came here in September, 1843, and remained until January. 'I was one of the waiters who attended upon them,' said Alfred. 'I used to stand on that rock and blow the horn to call them to dinner. There were fifteen of them, and they looked more like a company of skeletons than anything else.' One of the number died here. His case was hopeless when he entered, and even when conscious that his end was near he refused to leave. I can conceive of one man being benefited by a residence in the Cave, but the idea of a company of lank, cadaverous invalids wandering about in the awful gloom and silence, broken only by their hollow coughs, doubly hollow and sepulchral there, is terrible. On a mound of earth near the Dining Room I saw some cedar trees, which had been planted there as an experiment. They were entirely dead, but the experiment can hardly be considered final, as the cedar is, of all trees, the most easily injured by being transplanted."

It is surprising that such an observing traveler as Mr. Taylor should have fallen into so palpable an error as to imagine that trees, or any other species of vegetation, could possibly maintain vitality under circumstances where light, moisture, and heat are absent. This part of the Cave is perfectly dry; but the want of light would alone be sufficient to prevent, or destroy vegetation.

Mr. Taylor continues: "I now noticed that the ceiling became darker, and that the gray cornice of the walls stood out from it in strong relief. Presently it became a sheet of unvarying blackness, which reflected no light, like a cloudy night sky All at once a few stars glimmered through the void, then more and more, and a firmament as far off and vast, apparently, as that which arches over the outer world, hung above our heads. We were in the celebrated Star Chamber. Leaning upon a rock which lay upon the right side of the avenue, we looked upward, lost in wonder at the marvelous illusion. It is impossible to describe the effect of this mock sky. Your reason vainly tells you that it is but a crust of black gypsum, sprinkled with points of the white limestone beneath, seventy five feet above your head. You see that it is a fathomless heaven, with its constellations twinkling in the illimitable space. You are no longer upon this earth. You are in a thunder riven gorge of the mountains of Jupiter, looking up at the strange firmament of that darker planet. You see other constellations rising, far up in the abyss of midnight, and witness the occultation of remoter stars."

The starry firmament which Mr. Taylor has so graphically described, could not be seen to advantage under ordinary circumstances, every visitor having his lamp in hand. The guide seats the visitors upon a bench provided for the purpose, placed against the right wall of the avenue; he then takes all the lamps from the party, and, stepping back ten or fifteen feet, on the same side, he holds them within a small recess naturally formed in the rock, in such a manner that none of the direct rays of the light fall upon the eyes of the beholder, but are thrown upon the ceiling. By this manoeuvre the illusion of a starry sky is as complete as it is possible to be; a perfect representation of a comet, as if especially provided to add to the reality of the sublime scene, is also plainly discoverable in the distance.

After indulging the visitors in the fascination of the scene long enough to produce a lasting impression, the guide, with the lamps, passes to the opposite side of the avenue, in front of us, leaving us seated as before, and descends into the mouth of an avenue still lower. As he slowly disappears below, we have the finest display of lights and shadows that it is possible to imagine. A black cloud gradually passes over the sky, and it is difficult to divest one's self of the idea that a storm is approaching. It needs but the flash of lightning and the roar of thunder to make the illusion complete.

After producing the storm illusion, the guide disappears entirely with the lamps through the nether avenue which communicates with the one by which we entered several hundred yards in the rear. We are thus left in total darkness, without even the sight of the midnight sky to console us. Of these moments, Bayard Taylor remarks: "Yes, this is darkness, solid, palpable darkness. Stretch out your hand and you can grasp it; open your mouth and it will choke you. Such must have been the primal chaos before Space was, or Form was, or 'Let there be light!' had been spoken. In the intense stillness I could hear the beating of my heart, and the humming sound made by the blood in its circulation."

After waiting a short time, sufficiently long to enable us to appreciate the sense of total darkness, we observed the faintest rays of daylight in the eastern horizon; and then, to heighten the illusion, we heard the well imitated crow of chanticleer. Day was breaking after that period of awful darkness; lighter and lighter came the morning as the guide slowly approached, for it was he, until finally his lights came in full view, giving us one of the finest artificial sunrises that could possibly be produced.

The sights and illusions of the Star Chamber are so wonderful and so complete, that when we reluctantly take our leave of it, we feel as though we had passed a night in a new world, and that the morning had unexpectedly broken upon us before our astonished faculties had had time to comprehend the extraordinary transition of time and circumstances.

With the exception of Echo River, says Dr. Wright, the Star Chamber is, perhaps, the most attractive object in the Cave.

The Floating Cloud Room connects the Star Chamber with Proctor's Arcade.

The clouds are produced by the scaling off of black gypsum from the ceiling, by an efflorescence of the sulphate of soda beneath it, by which a white surface is exposed. They appear to be drifting from the Star Chamber over the Chief City. The Cloud Room is a quarter of a mile in length, and in height and width corresponds with the Star Chamber.

# CHAPTER XVI. - PROCTOR'S ARCADE,

which is entered immediately beyond the Star Chamber, is, says Dr. Wright the most magnificent natural tunnel in the world. It is a hundred feet in width, forty five in height, and three quarters of a mile in length. The ceiling is smooth, and the walls vertical, and look as though they had been chiseled out of the solid rock.

When this tunnel is illuminated with a Bengal light at Kinney's Arena which is its western terminus, the view is magnificent beyond conception. This arcade is named in honor of Mr. L. J. Proctor, the proprietor of the Cave Hotel.

Kinney's Arena is a hundred feet in diameter and fifty feet in height. From the ceiling, in the center of the Arena there projects a stick, three feet in length and two inches in diameter. It rests parallel with the ceiling, and is inserted into a crevice in the rock. How it was placed in its present position is a difficult question to settle, inasmuch as it could not have been inserted in the position it occupies by artificial means.

After passing the S Bend, which has no particular points of attraction, Wright's Rotunda is entered.

This Rotunda is four hundred feet in its shortest diameter. The ceiling is from ten to forty five feet in height, and is perfectly level, the apparent difference in height being produced by the irregularity of the floor. It is astonishing that the ceiling has strength to sustain itself, for it is not more than fifty feet from the surface of the earth. Fortunately the Cave at this point is perfectly dry, and no change of any kind is transpiring in it, otherwise there might be some risk of it falling in, as evidences of such occurrences are to be found in the surrounding country.

When this immense area is illuminated at the two extremes simultaneously, it presents a most magnificent appearance.

At the eastern extremity of the Rotunda is a column, four feet in diameter, extending from the floor to the ceiling, termed Nicholas' Monument, after one of the old colored guides.

The Fox Avenue communicates with the Rotunda and S Bend. It is about five hundred yards in length, and is worth exploring.

A short distance beyond Wright's Rotunda the Main Cave sends off several avenues or branches. That to the left leads to the Black Chamber, which is one hundred and fifty feet wide and twenty in height, the walls and ceiling of which are incrusted with black gypsum. It is the most gloomy room in the Cave.

There are two avenues leading off to the right. The far one communicates with Fairy Grotto, which contains a most magnificent collection of stalagmites. It is a mile in length. The other avenue communicates with Solitary Cave, at the entrance of which there is a small cascade.

"We will at once enter the Fairy Grotto of the Solitary Cave. It is in truth a fairy grotto; a countless number of stalactites are seen extending, at irregular distances, from the roof to the floor, of various sizes and of the most fantastic shapes, some quite straight, some crooked, some large and hollow, forming irregularly fluted columns; and some solid near the ceiling, and divided lower down into a great number of small branches like the roots of trees, exhibiting the appearance of a coral grove. Hanging our lamps to the incrustations on the columns, the grove of stalactites became faintly lighted up, disclosing a scene of extraordinary wildness and beauty. 'This is nothing to what you will see on the other side of the rivers,' cries our guide, smiling at our enthusiastic admiration. With all its present beauty, this grotto is far from being what it was before it was despoiled and robbed some eight or nine years ago by a set of vandals, who, through sheer wantonness, broke many of the stalactites, leaving them strewn on the floor, a disgraceful memorial of their vulgar propensities and barbarian like conduct."

What is called the Chief City is situated in the Main Cave beyond the Rocky Pass.

It is about two hundred feet in diameter and forty in height. The floor is covered at different points with piles of rock, which present the appearance of the ruins of an ancient city.

Of this, and contiguous parts of the Cave, Bayard Taylor says: "Just one mile from the Star Chamber a rough stone cross has been erected to denote that the distance has been carefully measured. The floor here rises considerably which contracts the dimensions of the avenue, although they are still on a grand scale. About half a mile farther we come to the Great Crossings, where five avenues meet. In the dim light it resembled the interior of a great cathedral, whose arched roof is a hundred feet above its pavement. Turning to the left, at right angles to our former direction, we walked (still following the Main Avenue) some ten minutes farther, when the passage debouched into a spacious hall with a cascade pouring from the very summit of its lofty dome. Beyond and adjoining it was a second hall, of nearly equal dimensions, with another cascade falling from its roof. We turned again to the right, finding the avenue still more irregular and contracted than before, but had not advanced far before its ceiling began to rise, showing a long slope of loosely piled rocks lying in strong relief against a background of unfathomable darkness.

I climbed the rocks and sat down on the highest pinnacle, while Stephen descended to the opposite side of the slope, and kindled two or three Bengal lights, which he had saved for the occasion. It needed a stronger illumination than our lamps could afford to enable me to comprehend the stupendous dimensions of this grandest of underground chambers. I will give the figures, but they convey only a faint idea of its colossal character: length, eight hundred feet; breadth, three hundred feet; height, one hundred and twenty feet; area, between four and five acres. Martin's picture of Satan's Council Hall in Pandemonium would hardly seem exaggerated if offered as a representation of the Chief City, so far and vanishing is the perspective of its extremities, so tremendous the span of its gigantic dome.

"I sat upon the summit of the hill until the last fires had burned out, and the hall became even more vast and awful in the waning light of our lamps. Then taking a last look backward through the arch of the avenue, to my mind the most impressive view, we returned to the Hall of the Cascades. Stephen proposed showing me the Fairy Grotto, which was not far off; and to accomplish that end I performed a grievous amount of stooping and crawling in the S Cave. The Grotto, which is a delicate stalactite chamber, resembling a Gothic oratory, was very picturesque and elegant, and I did not regret the trouble I had taken to reach it."

To show the similarity of the impressions produced upon different individuals by these novel and remarkable sights, we quote from the "Visitor," who, in turn, quotes from Mr. Lee:

"Returning from the Fairy Grotto, we entered the Main Cave at the Cataract, and continued our walk to the Chief City, or Temple, which is thus described by Lee, in his Notes on the Mammoth Cave:

"The Temple is an immense vault covering an area of two acres, and covered by a single dome of solid rock, one hundred and twenty feet high. It excels in size the Cave of Staffa; and rivals the celebrated vault in the Grotto of Antiparos, which is said to be the largest in the world. In passing through from one end to the other, the dome appears to follow like the sky in passing from place to place on the earth. In the middle of the dome there is a large mound of rocks rising on one side, nearly to the top, very steep, and forming what is called the Mountain. When first I ascended this mound from the Cave below I was struck with a feeling of awe more deep and intense than anything that I had ever before experienced. I could only observe the narrow circle which was illuminated immediately around me; above and beyond was apparently an unlimited space, in which the ear could catch not the slightest sound, nor the eye find an object to rest upon. It was filled with silence and darkness; and yet I knew that I was beneath the earth, and that this space, however large it might be, was actually bounded by solid walls. My curiosity was rather excited than gratified. In order that I might see the whole in one connected view I built fires in many places with the pieces of cane that I found scattered among the rocks. Then taking my stand on the Mountain, a scene was presented of surprising magnificence. On the opposite side the strata of gray limestone, breaking up by steps from the bottom, could scarcely be discerned in the distance by the glimmering light. Above was the lofty dome, closed at the top by a smooth, oval slab, beautifully defined in the outline, from which the walls sloped away on the right and left into thick darkness. Every one has heard of the dome of the Mosque of St. Sophia, of St. Peter's and of St. Paul's; they are never spoken of but in terms of admiration, as the chief works of architecture, and among the noblest and most stupendous examples of what man can do when aided by science; and yet when compared with the dome of this Temple, they sink into comparative insignificance. Such is the surpassing grandeur of Nature's works."

"To us," adds "Visitor," "the Temple seemed to merit the glowing description above given; but what would Lee think, on being told that since the discovery of the rivers, and the world of beauties beyond them, not one person in fifty visits the Temple or the Fairy Grotto; they are now looked upon as tame and uninteresting."

From these justly merited descriptions of this portion of the Cave the reader may form some conception of the surpassing beauty and magnificence of other parts, when informed, as above, that the Chief City is now very rarely visited.

From the Chief City to the end of the Main Cave, a distance of three miles, there are several points at which the appearance which this avenue presented when filled with running water may be observed, where the overhanging cliffs closely resemble those in the Pass of El Ghor, of recent formation.

The Main Cave is terminated abruptly by rocks that have fallen from above. It must not, however, be supposed that this is the end of it, for there can be no doubt that it was closed at this point in the same manner as Dickson's Cave was terminated, and that the removal of the obstructing rock would open a communication with a cave of the same size as the one we have been attempting to describe.

Retracing our steps by the route just passed over, we now return to the Second Hoppers, used by the saltpetre miners, mount a flight of wooden steps, fifteen feet high, to the right of the Gothic Galleries, and enter the Gothic Arcade.

## CHAPTER XVII. - OF ANCIENT MUMMIES FOUND IN THE CAVE.

Upon ascending the ladder, and entering the Gothic Arcade, the first object to which our attention is directed is what is called the Seat of the Mummy, which consists of a niche in the left wall of the avenue, about forty yards from the steps, just large enough to accommodate a human being with a comfortable seat.

Dr. Wright informs us that the body of an Indian female was found in this niche, dressed in the skins of wild animals, and ornamented with the trinkets usually worn by the aborigines. We are also told, by the same authority, that, within a few feet distant, was at the same time discovered the body of an Indian child, attired in a similar manner, and in a sitting posture, resting against the wall. Both bodies are said to have been in a mummified condition. Dr. Wright suggests that they wandered into this avenue, and, becoming bewildered, sat down and died in the position in which they were found.

Dr. Wright does not state what became of the bodies, or any other particulars in addition to those we have just mentioned. We made inquiry of the guide as to what disposition had been made of the remains. He replied that he had no personal knowledge of the matter, as the discovery had been made long anterior to his Cave experience, but he had been informed that the bodies had been sent to Louisville, to some museum or medical college. We expressed our surprise that such extraordinary curiosities should have been removed from the place where found, where the changes of Time were unknown, or that they should have been carried beyond the Cave Hotel, at all events. The guide did not attempt any explanation. There are two niches in the side wall of solid rock, one about large enough for an adult to sit in, and the other of a size adapted to a child. When we come to reflect upon the subject, we are first astonished at the thought that a woman and young child should venture so far, without companions, into the dark cavern; and, as the entrance to the Gothic Arcade, from the Main Avenue, is effected by the aid of a ladder fifteen feet high, placed there since the discovery of the Cave by white men, we next wonder how it could have been possible for a woman and child to have made this difficult ascent without the aid of a ladder, and if possible, what object could have been in view sufficiently strong to have induced the woman with her infant to have surmounted such extraordinary obstacles? And, after overcoming all these difficulties, is it presumable that the woman and child should have continued to wander until they found two niches in the wall exactly adapted to their respective sizes, and they perhaps the only niches suited to the purpose in the Cave?

We are somewhat surprised that Dr. Wright, the latest recognized authority in caveology, does not enter more into detail respecting this important subject, particularly as several of his predecessors in Cave history have been much more minute in their accounts of the discovery and the disposition of the mummies. Not doubting Dr. Wright's conviction or the truth of the remarks that he has made upon the subject, we repeat that he, and others concerned, owe it as a duty to themselves to explain to the satisfaction of the public why the said bodies were removed from the position in which they were found (which, if remaining, would have constituted one of the greatest curiosities of the Cave), how they were disposed of; and in what condition they are at present.

In our researches relating to this interesting subject, we believe that we have found everything of importance that has yet been published; and with the design of laying before our readers the accounts that are at present accessible to but few, we hereby detail all that is known to us respecting the human mummies, and the bones of the lower animals, said to have been discovered in the Cave, that others may have the same data that we have upon which to found their conclusions.

The first record that we have regarding the mummies is to be found in a book entitled "The Hundred Wonders of the World, and of the Three Kingdoms of Nature, described according to the latest and best authorities, by the Rev. E. C. Clark," published at New Haven, 1821, which refers to the Mammoth Cave as one of the conspicuous wonders, under the title of "The Great Kentucky Cavern."

Many of what are now regarded as the chief wonders of the Cave, however, were totally unknown at that date (1821). The description is taken from an account given by Dr. Nahum Ward, which was published in the "Monthly Magazine," so far back as October, 1810. This account possesses but little value at the present day, except in the fact that it contains the earliest published notice, so far as we know, of the discovery of human remains within the Cave. In speaking of the mummy, Dr. Ward says (p. 109):

"It was removed from another cave for preservation, and was presented to him (Dr. W.), together with the apparel, jewels, music, etc. with which it was accompanied. It has since been placed in the Washington Museum, the proprietor of which thinks it probable that this mummy is as ancient as the immense mounds of the Western country, which have so much astonished the world."

No information is given us as to the location of the Washington Museum. This account does not contain anything further on the subject, than above quoted, which is much to be regretted.

Collins' Kentucky, a work containing much valuable information regarding the early history of the State, but which has become so scarce that we were unable to procure a copy until our manuscript was nearly ready for the press, in treating upon Edmonson County, gives a very entertaining account of the Mammoth Cave (pp. 254-61). The following is the opening paragraph:

"In Edmonson County is situated, perhaps the greatest natural wonder in the world, the celebrated Mammoth Cave. In no other place has nature exhibited her varied powers on a more imposing scale of grandeur and magnificence. The materials of the following sketch of this Cave are derived, principally, from a small publication issued by Morton &

Griswold, of Louisville, entitled Rambles in the Mammoth Cave, during the year 1844, by a Visitor. This publication contains, we believe, the most complete and accurate description of this subterranean palace that has yet appeared, and gives the reader a very vivid conception of that amazing profusion of grand, solemn, picturesque, and romantic scenery, which impresses every beholder with astonishment and awe, and attracts to this Cave crowds of visitors from every quarter of the world."

After speaking of the surroundings and of the entrance of the Cave, the author, referring, we suppose, to what is now known as the Rotunda, says:

"The entire extent of this prodigious space is covered by a single rock, in which the eye can detect no break or interruption, save at its borders, which are surrounded by a broad, sweeping cornice, traced in horizontal panel work, exceedingly noble and regular. Not a single pier or pillar of any kind contributes to support it. It needs no support; but is by its own weight made steadfast and immovable.

"At a very remote period," continues our author, "this chamber seems to have been used as a cemetery; and there have been disinterred many skeletons of gigantic dimensions, belonging to a race of people long since vanished from the earth. Such is the vestibule of the Mammoth Cave. The walls of this chamber are so dark that they reflect not a single ray of light from the dim torches."

This is the first intimation given by the writer of the discovery of human remains in the Cave, and is all that he says upon the subject in this place.

After proceeding, however, for some time, describing the various objects of interest upon entering the Gothic Arcade, he recurs to the subject; and, in consideration of the fact that we had expressed doubts regarding the whole question, before this account came under our notice, we here quote the entire remarks concerning the mummies, that our readers may decide for themselves what amount of credibility the account is entitled to:

"The Gothic Avenue, to which the visitor ascends from the main cave by a flight of stairs, is about forty feet wide, fifteen feet high, and two miles long. The ceiling in many places is as smooth and white as if formed by the trowel of the most skillful plasterer. In a recess on the left hand, elevated a few feet above the floor, two mummies, long since taken away, were to be seen in 1813. They were in good preservation, one was a female, with her extensive wardrobe placed before her. Two of the miners found a mummy in Audubon Avenue in 1814; but, having concealed it, it was not found until 1840, when it was so much injured and broken to pieces by the weights which had been placed upon it as to be of no value. There is no doubt that by proper efforts discoveries might be made which would throw light on the history of the early inhabitants of this continent. A highly scientific gentleman of New York, one of the early visitors to the Cave, says in his published narrative:

"'On my first visit to the Mammoth Cave in 1813, I saw a relic of ancient times, which requires a minute description. This description is from a memorandum made in the Cave at the time.

"In the digging of saltpeter earth in the Short Cave a flat rock was met with by the workmen, a little below the surface of the earth, in the Cave; this stone was raised, and was about four feet wide, and as many long; beneath it was a square excavation about three feet deep, and as many in length and width. In this small nether subterranean chamber sat in solemn silence one of the human species, a female, with her wardrobe and ornaments placed at her side. The body was in a state of perfect preservation, and sitting erect. The arms were folded up, and the hands were laid across the bosom; around the two wrists was wound a small cord, designed, probably, to keep them in the posture in which they were first placed; around the body and next thereto were wrapped two deer skins. These skins appeared to have been dressed in some mode different from what is now practiced by any people of whom I have any knowledge. The hair of the skins was cut off very near to the surface. The skins were ornamented with the imprints of vines and leaves, which were sketched with a substance perfectly white. Outside of these two skins was a large square sheet, which was either wove or knit. The fabric was the inner bark of a tree, which I judge from appearances to be that of the linn tree. In its texture and appearance it resembled the South Sea Island cloth or matting; this sheet enveloped the whole body and head. The hair on the head was cut off within an eighth of an inch of the skin, except near the neck, where it was an inch long. The color of the hair was a dark red; the teeth were white and perfect. I discovered no blemish upon the body, except a wound between two ribs, near the backbone; and one of the eyes had also been injured. The finger and toe nails were perfect and quite long. The features were regular. I measured the length of one of the bones of the arm with a string, from the elbow to the wrist joint, and they equaled my own in length, viz., ten and a half inches. From the examination of the whole frame I judged the figure to be that of a very tall female, say five feet ten inches in height. The body, at the time it was discovered, weighed but fourteen pounds, and was perfectly dry; on exposure to the atmosphere, it gained in weight, by absorbing dampness, four pounds. Many persons have expressed surprise that a human body of great size should weigh so little, as many human skeletons, of nothing but bone, exceed this weight.

"Recently experiments have been made in Paris which have demonstrated the fact of the human body being reduced to ten pounds by being exposed to a heated atmosphere for a long period of time. The color of the skin was dark, not black; the flesh was hard and dry upon the bones. At the side of the body lay a pair of moccasins, a knapsack, and an indispensable, or reticule. I will describe these in the order in which I have named them. The moccasins were made of wove or knit bark, like the wrapper I have described. Around the top was a border to add strength, and perhaps as an ornament. These were of middling size, denoting feet of a small size. The shape of the moccasins differs but little from the deer skin moccasins worn by the northern Indians. The knapsack was of wove or knit bark, with a deep, strong border around the top, and was about the size of knapsacks used by soldiers. The workmanship of it was neat, and such as would do credit, as a fabric, to a manufacturer of the present day. The reticule was also made of wove or knit bark. The shape was much like a horseman's valise, opening its whole length on the top. On the side of the opening, and a few inches from it, were two rows of loops, one row on each side. Two cords were fastened to the reticule at the top, which were passed through the loop on one side, and then on the other side, the whole length, by which it was laced up and secured. The edges of the top of the reticule were strengthened with deep, fancy borders. The articles contained in the knapsack and reticule were quite numerous, and were as follows: one head cap, made of wove or knit bark, without any border, and of the shape of the plainest night cap; seven head dresses, made of the quills of large birds, and put together somewhat in the way that feather fans are made, except that the pipes of the quills are not drawn to a point, but are spread out in straight lines with the top. This was done by perforating the pipe of the quill in two places, and running two cords through the holes, and then winding around the quills and the cord fine thread to fasten each quill in the place designed for it. These cords extended some length beyond the quills on each side, so that on placing the feathers erect the cords could be tied together at the back of the head. This would enable the wearer to present a beautiful display of feathers standing erect, and extending a distance above the head and entirely surrounding it. These were most splendid head dresses, and would be a magnificent ornament to the head of a female at the present day. Several hundred strings of beads; these consisted of very hard, brown seed, smaller than hemp seed, in each of which a small hole had been made, and through the whole a small three corded thread, similar in appearance and texture to seine twine; these were tied up in bunches, as a merchant ties up coral beads when he exposes them for sale. The red hoofs of fawns on a string supposed to be worn around the neck as a necklace. These hoofs were about twenty in number, and may have been emblematic of innocence. The claw of an eagle, with a hole made in it, through which a cord was passed, so that it could be worn pendant from the neck. The jaw of a bear, designed to be worn in the same manner as the eagle's claw, and supplied with a cord to suspend it around the neck. Two rattlesnake skins; one of these had fourteen rattles; these skins were neatly folded up. Some vegetable colors done up in leaves. A small bunch of deer sinews, resembling catgut in appearance. Several bunches of thread and twine, two and three threaded, some of which were nearly white. Seven needles, some of which were of horn and some of bone; they were smooth, and appeared to have been much used. These needles had each a knob or whorl on the top, and at the other end were brought to a point like a large sail needle. They had no eyelets to receive a thread. The top of one of these needles was handsomely scalloped. A hand piece made of deer skin, with a hole through it for the thumb, and designed probably to protect the hand in the use of the needle, the same as thimbles are now used. Two whistles, about eight inches long, made of cane, with a joint about one third the length; over the joint is an opening extending to each side of the tube of the whistle; these openings were about three quarters of an inch long and an inch wide, and had each a flat reed placed in the opening. These whistles were tied together with a cord wound around them.

"I have been thus minute in describing this mute witness from the days of other times, and the articles which were deposited within her earthly house. Of the race of people to whom she belonged when living we know nothing; and, as to conjecture, the reader who gathers from these pages this account can judge of the matter as well as those who saw the remnant of mortality in the subterranean chambers in which she was entombed. The cause of the preservation of her body, dress, and ornaments is no mystery. The dry atmosphere of the Cave, with the nitrate of lime with which the earth that covers the bottom of these nether palaces is so highly impregnated, preserves animal flesh, and it will neither putrefy nor decompose when confined to its unchanging action. Heat and moisture are both absent from the Cave, and it is these two agents acting together which produce both animal and vegetable decomposition and putrefaction.

"In the ornaments, etc. of this mute witness of ages gone we have a record of olden time, from which, in the absence of a written record, we may draw some conclusions. In the various articles which constituted her ornaments there were no metallic substances; in the make of her dress there is no evidence of the use of any other machinery than the bone and horn needles. The beads are of a substance of the use of which for such purposes we have no account among people of whom we have any written record. She had no warlike arms. By what process the hair of the head was cut short, or by what process the deer skins were shorn, we have no means of conjecture. These articles afford us the same means of judging of the nation to which she belonged, and of their advances in the arts, that future generations will have in the exhumation of a tenant of one of our modern tombs, with the funeral shroud, etc. in a state of like preservation, with this difference, that with the present inhabitants of this section of the globe but few articles of ornament are deposited with the body. The features of this ancient member of the human family much resembled those of a tall, handsome, American woman. The forehead was high, and the head well formed."

---

This constitutes what appears to be, in the estimation of the historian of Kentucky, the most valuable part of the history of the Mammoth Cave. The name of the writer is not given. It is simply stated that the account of this mummy was published "by a highly scientific gentleman of New York, one of the early visitors of the Cave."

In examining Collins' account of the Cave, from whom we had a right to expect a very full history, for he speaks of it as being "perhaps the greatest natural curiosity of the world," we were much disappointed to find that all reference to its early history was omitted. He does not state the year of the discovery, by whom it was discovered, or what led to its discovery. In this respect the letter we give from Mr. Gorin contains information that, to our knowledge, has not hitherto been published. Mr.

C. dwells at great length on what we regard as minor points of interest, while Echo River and the great curiosities beyond are scarcely mentioned, and no attempt made at description, the whole being summarily dismissed with the remark that "a detailed description of these wonders would not consist with the plan of this work." And this statement is made after the minute description of the mummies and their ornaments.

The work before mentioned, entitled "*The Universe; or, The Infinitely Great and the Infinitely Small,*" by F. A. Poucher, M.D., Corresponding Member of the Institute of France, etc., translated from the French, illustrated, and published by Charles Scribner & Co., New York, 1870, has recently fallen under our notice (March, 1870). The character of the work is not very definitely implied in the title; it is designed as a popular natural history, and treats of botany, zoology, ornithology, geology, etc. Under the latter head five or six pages are devoted to an account of the Mammoth Cave. Of the nature of the curiosity the author remarks:

"The Mammoth Cave of the United States owes its renown not to the celebrity of those who have visited it, but to its extent, which is perhaps greater than that of any other existing Cave."

Again he says: "The Mammoth Cave is always an object of great interest to the Americans. They go there in crowds, and there is not always accommodation in the great hotel intended to receive the tourists, although it is arranged for three hundred guests. The exploration requires five or six days, and an army of guides is always kept for the service of travelers."

In the above paragraph the author commits two errors: Americans do not go in crowds, as they should; and the enterprising and gentlemanly proprietor of the hotel, Mr. L. J. Proctor, has so far always been able to accommodate satisfactorily all the crowds that have presented themselves.

Farther on the author states that, "Up to the present time 226 avenues have been made out, besides 57 domes, 11 lakes, 7 rivers, 8 cataracts, and 32 abysses, some of which are of immense depth." Those readers who have followed us closely will recognize that the above estimate is considerably magnified, on the principle, we presume, that "distance lends enchantment to the view;" for it does not appear in the record that the writer had ever made a personal inspection of the things whereof he speaks.

Dr. Poucher speaks of the human mummies and human skeletons found in the Cave; also, of the discovery therein of the bones of the bear, hyena, and mastodon. We are not aware that he obtained this information from any American authority. Indeed, he does not cite authorities for any of his assertions regarding the Cave. Dr. P. mentions the blind fishes of the Cave, and gives a tolerably accurate diagram of them. He does not refer to the recognized name given them by Professor Agassiz *(Amblyopsis spelaeus),* but calls them *Cyprinodonts.* He adds that they appear to be devoid of eyes. This question, as our readers will have already learned, has long since been settled in this country. Dr. P. also alludes to the fact that the Cave has been (and he appears to think still is) used as a sanitarium, speaking of it as "the sulphurous atmosphere in which medical men kept their patients afflicted with chest affections." Of all affections those of the chest would be the most aggravated by a sulphurous atmosphere; and we are surprised that Dr. P., a medical man, should make such a statement without interposing an interjection of astonishment, and without giving his authority for it. There is no sulphurous atmosphere in any part of the Cave, though one or two sulphur springs are found.

In connection with Dr. Poucher's remarks upon the Mammoth Cave there is an illustration of a view of the Dead Sea, which is drawn entirely from fancy. There is also another very conspicuous illustration of the River Styx. In the scene a boat is represented, propelled by a negro standing in the prow, with a single oar or paddle, who is shown to be hatless and naked to the loins, and appears to be making frantic efforts to advance, as though hoping thereby to escape from some desperate pursuer. (The negro cannot endure as much cold as the white man, and it can scarcely be presumed that he would enjoy any amount of comfort in this condition at the temperature of 59°.) In the stern sits a white man holding in his hand, in front and above his head, a flaming torch, straining his eyes as if in expectation of discovering in the impenetrable darkness before him the forbidding gnomes of this nether world. This picture is in striking contrast with the reality. Instead of the wild excitement manifested in the countenances of the boatman and voyager, this passage is one of the calmest, most placid, and dreamy that can be imagined; it is a quiet, though grand, embarkation "over the smooth surface of the summer sea," where no fear is ever felt regarding the intrusion of evil spirits.

Dr. Poucher's notice of the Mammoth Cave, we are compelled to say, is full of errors: there is scarcely a paragraph that can be accepted as literally true.

The discovery of so many errors upon this subject (the Cave) in such a pretentious and expensive work as that of Dr. Poucher's *Universe,* at this late date (1870), the more clearly convinces us of the necessity for the publication of an historical narrative of the Cave which, while entering more fully into detail than any previous work on the subject, can be relied upon as accurate in its statements, so far as their nature is susceptible of positive demonstration.

The most recent, and best authenticated information we have been able to find upon the subject of the Cave Mummies, is contained in a letter from Mr. Proctor, the present proprietor of the hotel, dated March 12, 1870. He says (in reply to our inquiries):

"There was a mummy found in the Mammoth Cave, and one in Short Cave, a cave in the neighborhood of the Mammoth. The one in Mammoth Cave was found in the Gothic Avenue, in 1815, by Mr. Ward, of Marietta, Ohio, and was sent to the Antiquarian Society of Worcester, Massachusetts, and is now there, as I learn by a letter of the Secretary of the Society; but is in a dilapidated condition. The one found in the Short Cave was taken and placed in the Museum at Cincinnati, and was burnt with that establishment many years ago. I have in my possession a photograph of the one taken out of the Gothic Avenue by Mr. Ward, in 1815."

In concluding our remarks on this subject, we emphatically agree with the author of "Rambles of a Visitor," etc. when he says, "The removal of those mummies from the place in which they were found can be viewed as little less than sacrilege. There they had been, perhaps, for centuries, and there they ought to have been left." The author adds, "What has become of them I know not. One of them, it is said, was lost in the burning of the Cincinnati Museum. The wardrobe of the female was given to a Mr. Ward, of Massachusetts, who, I believe, presented it to the British Museum."

NOTE. An article of some length, entitled "Underground Territories of the United States," appeared in "The International Magazine of Literature and Science," published by Stringer & Townsend, New York, in 1852 (vol. v.). The writer remarks:

"In Virginia, New York, and other States, the caves of Weyer, Schoharie, and many that are less famous, but not inferior in beauty or grandeur, are well known to travelers; but the Mammoth Cave under Kentucky is world renowned; and such felon States as Naples might hide in it from the scorn of mankind."

It is stated that the paper was prepared "chiefly from a letter by Mrs. Child, a very full description of this eighth wonder of the world, illustrated by engravings from recent drawings made under the direction of the Rev. Horace Martin, who proposes soon to furnish for tourists an ample volume on the subject."

The writer speaks of the mummies found in the Cave, and adds, "I believe that one of these mummies is now in the British Museum."

Nearly all the materials of which this article was composed, including the illustrations, were subsequently incorporated in Mr. Martin's book, from which we have repeatedly quoted. The original, however, did not come under our notice until quite recently.

We will also state that, since our text has been in type, we have obtained a copy of "The Monthly Magazine; or, British Register; reprinted, with American Intelligence, in Boston, U.S." In the number for April, 1816, under the head of American Intelligence, is given a "Description of the Great Cave in Warren [now Edmonson] County, Kentucky. Extract of a letter from Dr. Nahum Ward, formerly of Shrewsbury, Massachusetts, now resident in the Western country, to his friend in Worcester, giving an account of an excursion in Kentucky in the fall of last year; dated Marietta, April 4, 1816."

This account appears to be copied from a newspaper called the "Worcester Spy;" and this is the original publication of the history of the mummy which we have extracted from "Collins' Kentucky." "The Monthly Magazine," however, did not copy the details relative to the mummy (dress, ornaments, etc.) which we have seen was so fully copied by Mr. Collins. We find nothing in this article the quotation of which would afford additional interest to our readers. In a subsequent number of the same magazine, July, 1816, Dr. Ward furnished a map of the Cave, together with an engraving of the mummy which was therein found, and with which he had been presented.

The map, of course, is chiefly drawn from imagination: at that early date no surveys had been attempted. Nearly all the names by which the various parts of the Cave were then described have, since that date, been changed.

As might naturally be expected in describing a curiosity so extraordinary in its dimensions and characteristics, and of which so little was known at that day, we find considerable exaggeration and some misstatements in the account of Dr. Ward. He speaks of various chambers that constitute an area of from six to eight acres, and estimates that he explored a continuous avenue to the distance of eleven miles. The "Bottomless Pit" was not crossed for more than twenty years afterward; and the extreme length of the "Long Route," now known, does not exceed nine miles. The writer also speaks of Green River as passing over several of the avenues of the Cave, the incorrectness of which statement has long since been established.

The drawing of the mummy, which accompanies Dr. Ward's map, represents much more faithfully the features and form of the male than the female. It is in the sitting position, with the arms folded across the breast, the position which it is represented to have occupied when discovered.

The map, the drawing of the mummy, and the account of the Cave, as furnished by Dr. Ward and Mrs. Child, are interesting to the student of the Cave history, but, at the present day, are really of no practical value; and we simply note the articles in these magazines to indicate to our readers the character of the field over which we have traveled in pursuing our design of making our work as complete as possible.

# CHAPTER XVIII. - INSTANCES OF PERSONS BECOMING LOST IN THE CAVE.

## THE PROPER COURSE TO PURSUE IN SUCH CASES.

It is said that a person lost in the Cave, without any hope of escape, would undoubtedly die in a very short time. That this is the case the history of those who have been lost in it would seem to prove.

Thus, on one occasion, says Dr. Wright, a gentleman wandered from his party, when by some accident his lamp was extinguished. In endeavoring to make his escape he became alarmed, and finally insane, and crawling behind a large rock, remained in that position for forty eight hours; and although the guides repeatedly passed the rock behind which he was secreted, in search for him, he did not make the slightest noise; and when finally discovered, endeavored to make his escape from them, but was too much exhausted to do so.

In another instance, we are told, a lady allowed her party to get so far in advance that their voices could no longer be heard, and in attempting to overtake them, fell and extinguished her lamp, when she became so terrified at her situation that she swooned; and when discovered a few minutes afterward, and restored, she was found to be in a state of insanity, from which she did not recover for a number of years.

The author of " Rambles," etc. quotes the following case from the author of "Calavar:"

"In the Lower Branch is a room called the Salts Room, which produces considerable quantities of the sulphate of magnesia, or of soda, we forget which, a mineral that the proprietor of the Cave did not fail to turn to account. The miner in question was a new, raw hand, of course neither very well acquainted with the Cave itself; nor with the approved modes of averting or repairing accidents, to which, from the nature of their occupation, the miners were greatly exposed. Having been sent, one day, in charge of an older workman, to the Salts Room to dig a few sacks of the salt, and finding that the path to this sequestered nook was perfectly plain; and that, from the Haunted Chambers being a single continuous passage without branches, it was impossible to wander from it, our hero disdained, on his second visit, to seek or accept assistance, and trudged off to his work alone. The circumstance being common enough, he was speedily forgotten by his brother miners; and it was not until several hours after, when they had left off their toil for the more agreeable duty of eating their dinner, that his absence was remarked, and his heroical resolution to make his way alone to the Salts Room remembered. As it was apparent, from the time he had been gone, that some accident must have happened to him, half a dozen men, most of them negroes, stripped half naked, their usual working costume, were sent to hunt him up; a task supposed to be of no great difficulty, unless he had fallen into a pit. In the meanwhile the poor miner, it seems, had succeeded in reaching the Salts Room, filling his sack, and retracing his steps half way back to the Grand Gallery; when, finding the distance greater than he thought it ought to be, the conceit entered his unlucky brain that he might, perhaps, be going wrong. No sooner had the suspicion struck him than he fell into a violent terror, dropped his sack, ran backward, then returned, then ran back again, each time more frightened and bewildered than before; until, at last, he ended his adventure by tumbling over a stone and extinguishing his lamp. Thus left in the dark, not knowing where to turn, frightened out of his wits besides, he fell to remembering his sins, always remembered by those who are lost in the Cave, and praying with all his might for succor. But hours passed away, and assistance came not; the poor fellow's frenzy increased; he felt himself a doomed man; he thought his terrible situation was a judgment imposed on him for his wickedness; nay, he even believed, at last, that he was no longer an inhabitant of the earth, that he had been translated, even in the body, to the place of torment; in other words, that he was in hell itself, the prey of the devils, who would presently be let loose upon him. It was at this moment that the miners in search of him made their appearance. They lighted upon his sack, lying where he had thrown it, and set up a great shout, which was the first intimation he had of their approach. He started up, and seeing them in the distance, the half naked negroes in advance, all swinging their torches aloft, he, not doubting that they were those identical devils whose appearance he had been expecting, took to his heels, yelling for mercy; nor did he stop, notwithstanding the calls of his amazed friends, until he had fallen a second time over the rocks, where he lay on his face, roaring for pity, until, by dint of much pulling and shaking, he was convinced that he was still in the world and the Mammoth Cave."

Such is the story of the Haunted Chambers, the name having been given to commemorate this incident.

Not a year passes, we are informed, but the guides have to go in search of persons who have been foolhardy enough to leave their party, and who, in every instance, become speedily bewildered, and when discovered are in the act of crying or at prayer. In such cases the guides are overpowered with kisses, embraces, and other demonstrations of gratitude.

The proper course for persons to pursue when lost in the Cave is for them to remain in the place where they first became confused, and not to stir from it until rescued by the guides. They will not have to wait more than from three to ten hours from the time at which they should have returned to the hotel.

# CHAPTER XIX. - GOTHIC ARCADE.

The Gothic Arcade is entered by ascending a flight of steps, as before mentioned; and, after passing the seat of the Mummy a short distance, there is to be found a large stalactite, which extends from the floor to the ceiling, termed the Post Oak, from its fancied resemblance to a variety of oak tree that grows in the vicinity of the Cave.

The First Echo is the name given to that part of Gothic Arcade which passes over Pensacola Avenue, the floor of which, when forcibly struck, emits a hollow sound. This hollow sound is observed even in walking, conveying the impression that some danger of falling through might result from too heavy a tread.

For more than a quarter of a mile this avenue has a ceiling perfectly flat, with every appearance of having received a coat of plaster. It is smoked over in all parts with the names of vulgar visitors, from which circumstance one locality is called the Register Room. Persons formerly carried candles in their trips through the Cave, and, by tying them to poles, succeeded in not only smoking their names upon the ceiling, from eight to sixteen feet overhead, but in many instances their portraits, for there were frequently rude attempts at drawing the figures of sheep and pigs, as we are told by Bayard Taylor, and as every visitor may see for himself. The lamps now in use are much more convenient for carrying, and have the additional advantage of guarding against all such desecration.

The Register Room, to use the words of Dr. Wright, is about three hundred feet long, forty wide, and from eight to sixteen in height. The ceiling is white, and as smooth as though it had been plastered. In this room hundreds of persons have displayed their bankruptcy in everything pertaining to good breeding and taste by tracing their obscure names on the ceiling with the smoke of a candle.

After passing the Register Room, the ceiling gradually becomes broken and rugged, studded here and there with unfinished stalactites.

The next point of interest reached is the Gothic Chapel. This is a large room, the ceiling of which appears to be supported by gigantic stalactites, which extend to the floor. These stony icicles become large enough to form ribbed pillars and fair Gothic arches. When a number of lamps are hung upon these columns, this room presents a beautiful appearance.

We are informed that a romantic marriage once took place in this chapel, which family interference prevented occurring on the earth. It is said that the fair lady, whose lover was opposed by her parents, in a rash moment promised them that she would never marry her betrothed on the face of the earth. Afterward, repenting her promise, but being unable to retract, and unwilling to violate it, she fulfilled her vow to her parents, as well as to her lover, by marrying him under the earth. This is but another illustration of the proverbially ingenious management of woman.

*Gothic Chapel*

*The Altar*

Two of the stalactites in this chapel are called the Pillars of Hercules, which are said to be thirty feet in circumference; and we are told that, in the formation of a stalactite, a period of fifty years is required to produce an incrustation of the thickness of a wafer. We have not attempted to calculate the time necessary, at this rate, to form a stalactite of the size named. Geologists, in estimating the antiquity of the origin of the earth, might find important data here that, to our knowledge, has not been heretofore used. But it would be impossible to approximate accuracy in such calculations, from the fact that it may have been, for all we know, twenty thousand years since the cessation of the process of formation, for this appears to be one of the oldest avenues in the Cave.

To the ceiling of the Chapel are attached great numbers of small stalactites, from six to eighteen inches in length, giving a very singular and handsome view overhead. The Gothic Chapel has been very accurately and strikingly photographed by the parties before Mentioned.

Speaking of the immense stalactites found in Gothic Chapel, reminds us that we have not explained the meaning of the terms "stalactite" and "stalagmite," which formations being peculiar to caves, it is presumed that the majority of readers are not familiar with the process of their formation. We will explain the process in the language of Dr. Wright, which we do not think can be improved upon either in point of fact or brevity:

"When water, holding the bicarbonate of lime in solution, drops slowly from the ceiling, by which it is exposed to the air sufficiently long to allow the escape of one equivalent of carbonic acid gas, the lime is deposited in the form of the proto-carbonate of lime. If the deposit occurs in such a manner that the accumulation takes place from above downward, in the form of an icicle, it constitutes what is termed a stalactite; but if it accumulates from below upward, it is called a stalagmite. Stalactites and stalagmites frequently meet in the center and become cemented, by which a column of support is formed."

If the limestone which forms the stalactite is perfectly pure, it will be white, or semi-transparent; if it contains oxide of iron, it will be of a red or yellowish color. The black stalactites contain a large proportion of the black oxide of iron.

Leaving Gothic Chapel, and pursuing our course, we are next introduced to Vulcan's Smithy, a room the floor of which is strewn with stalagmitic nodules, colored with the black oxide of iron, which resemble the cinders of a blacksmith's shop.

Bonaparte's Breastworks, immediately beyond, consist of a ledge of rocks that have been detached from the side of the avenue against which they rest.

The Arm Chair, called by the guides the Devil's Arm Chair, is the next object of interest. It is formed by the union of stalagmites and stalactites. It is told that the celebrated Jenny Lind rested in this Chair for some time during her visit to the Cave. Unless we make the statement ourself, we fear it will not be handed down to posterity, that we also rested for a few moments in the same regal Chair! The Chair is rather high for convenience; but still it accommodates an individual very comfortably.

The Elephant's Head is a large stalagmite, projecting from the left wall of the avenue, which is supposed to bear a striking resemblance to the head of the animal for which it is named.

A rock projecting sixteen feet over a pit which is seventy feet in depth, is fancifully denominated the Lover's Leap. It is not recorded, however, that any lover ever regarded his case so extremely desperate as to have induced him to take this fearful "leap in the dark."

After passing down a very precipitous bank, some thirty or forty feet on the left side of Lover's Leap, we enter a narrow avenue, very properly termed the Elbow Crevice. This avenue, though not more than from three to five feet wide, is fifty feet high, and twenty feet in length. It is another Fat Man's Misery, on an enlarged scale.

Gatewood's Dining Table is a flat rock which has been detached from the ceiling. It is about twelve feet long and eight wide, and is named after one of the former proprietors.

Next we reach Napoleon's Dome, which is fifty feet high, and from twenty to thirty wide. It was formed in the same manner as, and very much resembles, Corinna's Dome, in the Pass of El Ghor.

A pool of perfectly transparent water, called Lake Purity, is situated directly under Vulcan's Smithy; large amounts of the cinder like formations are to be seen in and around the miniature lake. Many small stalagmites are found in the bottom of the pool, which is quite shallow.

By the aid of the guide we succeeded in breaking off and securing two or three small specimens of the stalagmites from the bottom of Lake Purity. The Gothic Arcade terminates half a mile beyond Lake Purity, in a dome and small cascade; but visitors are rarely conducted beyond the lake.

This was the final end of our explorations in the Mammoth Cave. We retraced our steps to the Main Avenue, and wended our way to the mouth of the Cave. At the time of our exit on this occasion the daylight still prevailed, and the bright sunshine, besides being painful to the retina, presented a most singular appearance to our temporarily unaccustomed sight. Upon reaching the entrance, and looking out from behind the falling skein of water, the trees seemed to be illuminated with an unnatural fire. The daylight had a warm yellow line, intensely bright, and the sky was paler, but more luminous than usual. The air, by contrast with the exhilarating and pure atmosphere of the Cave, felt close, unpleasantly warm, and oppressive, like that of an ill ventilated greenhouse in winter. There was too much perfume in it, too many varieties of vegetable smells; for a short absence in the Cave, as before remarked, produces great acuteness of the olfactory nerves. A few minutes only, however, are required, after leaving the Cave, to enable the senses of sight and smell to resume their normal conditions.

*Devil's Arm Chair*

# CHAPTER XX. - SANITARY INFLUENCES OF THE CAVE.

Before quitting the Cave, we will refer to its sanitary influences. This was an interesting question some years ago. It has been asked what diseases are benefited, and what diseases are aggravated, by a brief Cave residence.

Persons afflicted with pulmonary consumption at one time resorted to the Cave for the benefit of its pure air and uniform temperature, in the vain hope of recovery. Several of them died there, and all of them succumbed soon after exposure to the external air. One patient did not see the light of the sun for a period of five months.

Several cottages, previously spoken of, built over twenty five years ago, for the residence of consumptives, at the entrance of Audubon's Avenue, and within the Gothic Arcade, are still to be seen. The idea that consumptive patients could be cured by a residence in the Cave must have resulted from a total misconception of the nature of the disease, as it is well known to the medical profession that the absence of light will develop the scrofulous diathesis, and cause a deposit of tubercles in the lungs.

The truth of this position was established in the cases of those who resorted to the Cave for relief; and the majority of those who remained any considerable length of time died within periods varying from three days to three weeks after leaving it. Those patients who remained in the Cave three or four months presented a frightful appearance. The face was entirely bloodless, eyes sunken, and pupils dilated to such a degree that the iris ceased to be visible, so that, no matter what the original color of the eye might have been, it soon appeared black.

Very few diseases, not even consumption, are aggravated by short and easy trips in the Cave. Chronic dysentery and diarrhea are said to have been cured by a short visit to the Cave, after all the usual remedies had failed.

In those diseases in which absolute silence and the total exclusion of light are indicated, the Cave, above all other places, possesses preeminent advantages; for nowhere else have we these conditions combined. Hence it may be inferred that brain affections, abnormal excitement of the brain, incipient insanity, etc., would undoubtedly be benefited by a temporary Cave residence. But practically we cannot assign any sanative virtues to a residence in the Cave. Too many conditions are absent that are necessary for the comfort and happiness of the patient. Dr. Wright remarks: "The only condition in which risk is incurred is during the menstrual period. Serious, and even fatal, results have been the consequence of inattention to this fact." No reason is assigned for this assertion, and we are unable to conjecture any; on the contrary, one or two instances have come under our notice where no bad effects resulted.

It is surprising how rapidly the quieting influence is felt in the Cave, it being indicated by pallor of the cheeks, yawning, and an almost irresistible tendency to sleep. Upon the first visit to the Cave this disposition is not so strongly manifested, for the reason that the attention is so constantly attracted by the novelty of the situation, and the ever changing and extraordinary sights. This tendency to sleep is not due to any impurity of the atmosphere, for, as already stated, it contains less carbonic acid than the outer air, but is referable solely to the complete silence and total absence of light.

Owing to the purity of the atmosphere and the even temperature, even delicate persons are enabled to take a much greater amount of physical exercise in the Cave than without. It is not an uncommon occurrence for an individual in delicate health to accomplish a journey of eighteen or twenty miles in the Cave, without suffering unusually from fatigue, who could not be prevailed upon to walk a distance of three miles on the surface of the earth.

After having accomplished our first day's journey in the Cave, we remarked to one of the gentlemen connected with the hotel, that we supposed ladies must suffer extremely from fatigue in going through the Long Route. He replied that such was not the case; and stated that, as a general rule, ladies endure the journey much better than men; and added that it was not an uncommon occurrence for ladies, after coming out in the evening, from a walk of eighteen miles, to enter the ballroom and dance until two o'clock in the morning!

# CHAPTER XXI. - PARTING REFLECTIONS.

We now take our leave of that dark, "mysterious realm," a wiser, if not a better man.

In the language of Professor Silliman, in the same article before quoted from, "I wish all my scientific friends could visit the Mammoth Cave; it teaches many lessons in a manner not to be learned so well elsewhere, and in this respect I was agreeably disappointed. I had heard that its interest was chiefly scenic; but I found it to exceed my utmost expectations as well in its illustrations of geological truth as in the wonderful character of its features. I will not detain you," he continues, "with any attempts at description of single parts, as no description can awaken those peculiar and deep emotions which a personal study of its details is calculated to produce."

In closing our narrative of the Cave we cannot more appropriately conclude than by giving the farewell words of Bayard Taylor. He expresses the same idea that we gave to the public in a brief newspaper notice ("Mobile Register," June, 1867) immediately after our visit, and before his sketch had come under our notice, and almost in the same language. His corroboration, together with that of Professor Silliman, gives additional weight to the remarks we made at that time regarding the character and magnitude of the curiosity. Mr. Taylor says:

"Before taking a final leave of the Mammoth Cave, let me assure those who have followed me through it, that no description can do justice to its sublimity, or present a fair picture of its manifold wonders. It is the greatest natural curiosity," adds this great traveler, "that I have ever visited, Niagara not excepted; and he whose expectations are not satisfied by its marvelous avenues, domes, and sparry grottoes, must either be a fool or a demi-god."

Whoever has seen a cascade, however diminutive in volume, can readily imagine a larger one, and by a greater effort of imagination, may conceive of the magnitude of Niagara; but he who has not entered the Mammoth Cave can form but faint conception of its character, or of the varied and lasting impressions produced upon the soul of him who has been wafted over its beautiful rivers, and whose spellbound steps have traversed its dark labyrinths, its vineyards, and its ever blooming floral bowers. Such scenes go with us in after days, and parting is truly a "vain adieu."

"Adieu to thee again! a vain adieu!

There can be no farewell to scenes like thine,

The mind is color'd by thy every hue."

In terminating this narrative of the Mammoth Cave, which has thus long, pleasantly and agreeably occupied our own thoughts, and profitably, we trust, the thoughts of our readers, in parting from those who now "have traced the Pilgrim to the scene which is his last," in this labor of love, we cannot, upon laying down our pen, give more fitting expression to our feelings than in the oft repeated, and oft-to-be-repeated words of the peerless Byron:

"Farewell! a word that must be, and hath been, A sound which makes us linger; yet, farewell!"

# DIAMOND CAVE.

Diamond Cave is an object of interest, and from its proximity to Mammoth Cave is worthy of mention in this connection, and should be visited by all curiosity seekers who may be attracted to the neighborhood by the great cavern, the account of which we have just completed.

This Cave is situated in Barren County (Ky.), one mile and a half from Glasgow Junction ("Bell's Tavern"), on the Louisville & Nashville Railroad, and five miles and a half from Mammoth Cave, being immediately on the road leading from the latter place to Glasgow Junction.

The first exploration of Diamond Cave, we are informed by its present proprietor, Mr. John R. Proctor. was made in July, 1859, by Dr. J. T. Andrews, of Montgomery, Ala.; Prof. T. A. Richardson, M.D., of New Orleans; Theodore H. Lowe, Esq., of Louisville; Mr. John Bell, proprietor of the well known " Bell's Tavern;" George Bliss, Esq., of New York; and Prof. C. W. Wright, of Louisville (author of the Guide Book of Mammoth Cave).

Mr. Proctor informs us that near this Cave there is another, rivaling it in beauty, which he explored in 1866; and, from the fact that many of the avenues are closed with masses of stalactites, he feels confident that a communication can be effected between the two. The stalactite and stalagmite formations in Diamond Cave are more numerous and far more beautiful than any found in Mammoth Cave. An English traveler remarked that, after many years' travel through Europe and America in search of geological and other objects of interest, he had not met with such a gem as Diamond Cave.

The following is a brief outline of the principal objects of interest to be found in this Cave:

In the Rotunda, which is seventy feet in diameter and thirty feet high, are to be seen Cleopatra's Needle, a stalagmite, five feet high and six inches in diameter, incrusted with the oxide of iron; Serpent's Head, directly over the Needle, five feet in length, depending from the ceiling, and hearing a striking resemblance to the head of a large snake, with its mouth open; Closed Lily is suspended from the ceiling, and closely resembles the flower after which it is named: it is eight feet long and two feet in diameter; Elfin's Grotto is a lovely alcove, fifteen feet above the floor of the Rotunda, the entrance of which is ornamented with beautiful stalagmites and stalactites: the crystallizations within assume every imaginable shape, among which can be seen a perfect resemblance to a cascade; Mammoth Stalagmite is eighteen feet high and seventy feet in circumference at its base, being by far the largest stalagmite known in the world.

It is stated that here, as in other caves, human bones were found in abundance at the base of the Armadillo, or Fallen Tree. They are said to have been discovered by Dr. Andrews, and that many were incrusted with the carbonate of lime.

After leaving the Rotunda we enter Lowe's Avenue, which varies from six to forty feet in height and from ten to forty in width. In this avenue are to be seen Stella Grotto, Vermiculated Ceiling, and many other objects of interest and beauty. After leaving the avenue we come to Andrews' Cascade, which is, without doubt, the most singularly beautiful formation of the kind ever discovered, exhibiting, when viewed from its base, a perfect resemblance to a waterfall, twenty eight feet in height and eighteen feet wide.

In Wright's Avenue, which extends from the Cascade to Diamond Grotto, are many tubular stalactites, which emit musical sounds when struck. Beyond Cascade Hall is seen the Magnolia Flower, a colossal flower six feet long and four feet in diameter, suspended twenty feet above the floor. It is composed of stalactitic plates of calcareous spar, and presents a perfect resemblance in form to the grande fleur of the Southern States. There are other formations near this of like character; and near Columbian Column there is a perfect representation of an immense Chandelier. From here a spar pavement extends, with slight interruptions, to the end of the Cave. It is composed of crystals of calcareous spar, which sparkle with great brilliancy as the lamps are moved to and fro above them. Next comes the Oriental Crystal Fans. From here to Fink's Acute Angle the avenue is really grand, having the appearance of white chalk cliffs.

Talia's Grotto is entered on the left of this avenue opposite the Atlantic Steamer. This is regarded as the most beautiful grotto in the world. The stalactites here are of the purest white, and rival in beautiful symmetry the finest Grecian carving. The Curtain Stalactite hangs upon the walls, which, from their peculiar tint, are called Blush Walls. It extends from the ceiling to within a few feet of the floor, and is so translucent that, by holding a light on the opposite side, the examiner can see through it. When struck it emits musical sounds.

The Pope is a stalagmite about ten feet in height and five feet in diameter at its base. It is composed of light stalagmitic marble, the surface of which reflects with great distinctness.

The Curtain Galleries and General Scott's Marquee cannot be excelled in beauty. Immediately to the left of the latter there is a miniature representation of Niagara Falls in winter.

Columbian Column is thirty feet high and ten feet in diameter at its base. It is most beautifully ornamented with fluted columns, ogees, cornices, and mouldings; and the entire surface displays innumerable crystals which sparkle on the approach of light like countless diamonds. We scarcely have any hesitation in saying that this particular point exhibits the most wonderfully beautiful formations of any to be found elsewhere; and this view, of itself, will more than repay the visitor for the time occupied in exploring Diamond Cave.

To the left of the Column is a collection of water called the Fountain of Orpheus. Bunker Hill Monument is a stalactite six feet high and six inches in diameter. Coral Pillar is a stalagmite four feet high and four inches in diameter, the surface of which is studded with little prominences resembling madrepore coral. A number of columns, similarly incrusted, are to be seen under the base of Columbian Column.

Eviscerated Body is to the left of the steps leading to Lover's Bower. When light is transmitted through it, it is a perfect representation of the body of a man from which the lungs have been removed.

Ameda Grotto has within it a pool of water lined with crystals, and encircled with stalagmitic formations. The stalactites in this grotto are of delicate form and of great variety.

Lot's Wife is a stalagmite four feet high, representing a figure draped in white. There is also to be found here an excellent representation of a Church Organ.

Diamond Grotto is twenty five feet in diameter and eight feet in height. The floor is covered with crystalline plates of calcareous spar, which sparkle with a brilliancy almost rivaling the gem for which the Cave is named. The ceiling is vermiculated, like the ceiling in Lowe's Avenue.

Among other places of interest in this beautiful Cave may be mentioned Bell's Spring, The Grand Retreat, Nettie's Palace, Mason Grotto, etc., which cannot here be more fully described.

All who visit Mammoth Cave should, before leaving the neighborhood, see the beauties of Diamond Cave.

## PROCTOR'S CAVE.

Proctor's Cave, the property of Mr. L. J. Proctor, is quite as remarkable for its beauty as Diamond Cave; and as our object is to give information to those desiring it, not only in relation to Mammoth Cave, but regarding the important curiosities in the immediate vicinity, our work would be incomplete without a special notice of Proctor's Cave.

This Cave is situated three miles from Mammoth Cave, and four miles from Glasgow Station, on the Louisville & Nashville Railroad. The picturesque scenery surrounding, the entrance of this Cave; the wonderful succession of domes, many of which are viewed from the base; the endless variety of stalactite and stalagmite formations, all contribute toward rendering a visit peculiarly attractive and interesting. Some even go so far as to say that there is not another Cave on this continent in which there is such a magnificent display of the chemical and mechanical action of water. The gypsum formations, such as rosettes, fibers, etc., are not less attractive. Dr. Wright tells us that the largest dome to be found in any Cave is to be found in Proctor's Cave. "It is at least three times as wide and long as Mammoth Dome in Mammoth Cave, and not less than one hundred and forty feet in height. The Curtain Dome is not less wonderful. Vast sheets of stalactite, yards in length and less than an inch in thickness, are arranged in the form of curtains, scrolls, etc. in endless variety."

Mammoth Cave is deficient only in stalactite formations. A visit to Proctor's and Diamond Caves renders cave knowledge and experience complete.

About three miles of this cave is now open to the public, through the greater part of which there is a substantial plank walk.

In a recent letter from Mr. Proctor to the author, he remarks, in speaking of this Cave, "I am constantly making new discoveries in this Cave; among others a large river, as yet inaccessible to visitors, and many beautiful domes and grottoes; and the probabilities are that it will prove to be a very large, as well as a beautiful Cave."

Glasgow Station is the site of what was for many years renowned as "Bell's Tavern," the stopping place heretofore for all Cave visitors; and was noted as being kept in a style superior to any country hotel in America. This, building was burned in 1860; and hence Cave City took the start. The property is now owned by Mr. Proctor, of Proctor's Cave, who is also the proprietor of the Mammoth Cave Hotel. He has a comfortable sized hotel now at Glasgow Junction, equal in capacity to the hotel at Cave City; and he informs us that he has laid the foundations of a first class stone hotel, and has the basement and second stories already up. The size is one hundred and twenty by sixty feet; it is to have sixty rooms, and will be completed during the present year (1870). Mr. Proctor promises to sustain the ancient reputation of "Bell's Tavern."

For the still further accommodation of Cave visitors, Mr. Proctor informs us that a charter for a railroad has been granted from Glasgow Junction to Brownsville, the county seat, and thence to intersect with the Elizabethtown & Paducah Railroad, which will pass directly by Diamond Cave, and Proctor's Cave, and within two and a half miles of Mammoth Cave. This road is expected to be completed about the middle of 1871. At present, stagecoaches run from Glasgow Station to Mammoth Cave, stopping at Diamond and Proctor's Caves, at a charge of one dollar per passenger.

Thus it will appear that, by this arrangement, visitors have offered to them the very desirable opportunity of seeing these three remarkable Caves with but little additional cost of time or money over that usually required in visiting the Mammoth Cave.

THE END.

www.ingramcontent.com/pod-product-compliance
Lightning Source LLC
Chambersburg PA
CBHW051515170526
45165CB00002B/485